AF333209

Are Wild Birds Important in the Transport of Arthropod-borne Viruses?

Ornithological Monographs

Editor: Michael L. Morrison

Department of Wildlife and Fisheries Sciences
210 Nagle Hall, 2258 TAMU
Texas A&M University
College Station, Texas 77843-2258

Managing Editor: Mark C. Penrose
Copy Editor: Richard D. Earles

Spanish translation of abstracts: Carlos Daniel Cadena,
Cintia Cornelius, and Lucio R. Malizia

The *Ornithological Monographs* series, published by the American Ornithologists' Union, has been established for major papers and presentations too long for inclusion in the Union's journal, *The Auk*.

Copying and permissions notice: Authorization to copy article content beyond fair use (as specified in Sections 107 and 108 of the U.S. Copyright Law) for internal or personal use, or the internal or personal use of specific clients, is granted by The Regents of the University of California on behalf of the American Ornithologists' Union for libraries and other users, provided that they are registered with and pay the specified fee through the Copyright Clearance Center (CCC), www.copyright.com. To reach the CCC's Customer Service Department, phone 978-750-8400 or write to info@copyright.com. For permission to distribute electronically, republish, resell, or repurpose material, and to purchase article offprints, use the CCC's Rightslink service, available on JSTOR at http://www.jstor.org/r/ucal/om. Submit all other permissions and licensing inquiries through University of California Press's Rights and Permissions website, www.ucpressjournals.com/reprintInfo.asp, or via e-mail: journalspermissions@ucpress.edu.

Back issues of *Ornithological Monographs* from earlier than 2007 are available from Buteo Books at http://www.buteobooks.com. Issues from 2008 forward are available from University of California Press, http://www.ucpressjournals.com. For complete abstracting and indexing coverage for *Ornithological Monographs*, please visit http://www.ucpressjournals.com.

Library of Congress Control Number: 2011905143
ISBN: 978-0-943610-90-0
Issued 15 July 2011
Ornithological Monographs, No. 71 viii + 64 pp.

Cover: A migratory flock of Cliff Swallows (*Petrochelidon pyrrhonota*) in western Nebraska en route to the species' wintering area in southern South America. Such movements of migratory birds have long been thought to be an important way that arthropod-borne viruses are transported across geographic regions, but direct evidence that migratory birds move these viruses is limited. Photograph by Charles R. Brown.

Printed by Allen Press on Forest Stewardship Council™–certified paper.

© 2011 American Ornithologists' Union. All rights reserved.

Are Wild Birds Important in the Transport of Arthropod-borne Viruses?

BY

CHARLES R. BROWN[1] AND VALERIE A. O'BRIEN[1,2]

[1]*Department of Biological Sciences, University of Tulsa, Tulsa, Oklahoma 74104, USA; and*
[2]*Department of Entomology and Plant Pathology, Oklahoma State University, Stillwater, Oklahoma 74078, USA*

ORNITHOLOGICAL MONOGRAPHS NO. 71

PUBLISHED BY
THE AMERICAN ORNITHOLOGISTS' UNION
WASHINGTON, D.C.
2011

This monograph is dedicated to William K. Reisen,
in recognition of his tireless efforts for over 40 years
to unravel the complexities of arbovirus transmission
and amplification.

TABLE OF CONTENTS

LIST OF TABLES

LIST OF FIGURES

From the Editor

I will readily admit that—at least prior to the submission of this monograph—I had only a general understanding of animal diseases. As an animal ecologist I certainly understood the substantial role that disease could have in local population dynamics. My attention to details of specific diseases was focused, however, on the diseases that affected the species I was currently studying; on those diseases I had to become better schooled. I am confident that we are all aware of the impact that diseases such as West Nile have had on certain bird populations. Given that I often pretend that I am a mammalogist and study rodents, I have long been aware of how I was playing with potential death when taking blood samples back before (i.e., pre-1990 or so) we knew about Hantavirus. But beyond a general knowledge, I was really unaware of how many diseases were lurking about us in the wild and how they were being transmitted.

Fortunately, we now have this detailed investigation by Charles R. Brown and Valerie A. O'Brien that takes us through the various theories about how birds might disperse certain diseases. As reviewed by the authors, the movements of migratory birds have been suggested as a major way in which viruses can be moved on local, continental, and even intercontinental scales. I certainly will not ruin the story line by summarizing what they found, but the story they tell is akin to a detective sorting through clues of highly different nature.

The conservation implications of the work by Brown and O'Brien go far beyond the direct effects that disease can have on bird populations. Indeed, if it were concluded (rightly or wrongly) that birds dispersed disease that affected human populations, then public support in general for many conservation activities aimed at birds could be negatively influenced. And as noted by the authors, false assumptions about the role of birds in dispersing pathogens could lead to a misdirection of public health resources into ineffective ways to predict or prevent disease spread.

This monograph summarizes a substantial body of knowledge, dispels certain false notions about the role of birds in the spread of pathogens, and provides a roadmap for future investigations. I, for one, am now much better informed about how we need to look at birds and their role in the spread of pathogens and disease in general.

Michael L. Morrison

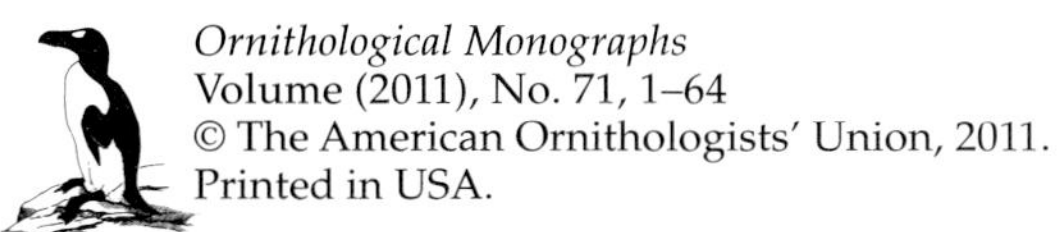

Ornithological Monographs
Volume (2011), No. 71, 1–64
© The American Ornithologists' Union, 2011.
Printed in USA.

ARE WILD BIRDS IMPORTANT IN THE TRANSPORT OF ARTHROPOD-BORNE VIRUSES?

CHARLES R. BROWN[1] AND VALERIE A. O'BRIEN[2]

Department of Biological Sciences, University of Tulsa, Tulsa, Oklahoma 74104, USA

ABSTRACT.—The encephalitic arthropod-borne viruses (arboviruses) can cause a variety of serious human and wildlife diseases, including eastern equine encephalomyelitis, western equine encephalomyelitis, St. Louis encephalitis, Japanese encephalitis, and West Nile neuroinvasive disease. Understanding how these pathogens are dispersed through the environment is important both in managing their health-related impact and in interpreting patterns of their genetic variability over wide areas. Because many arboviruses infect wild birds and can be amplified to a level that makes birds infectious to insect vectors, numerous workers have suggested that the movements of migratory birds represent a major way that these viruses can be transported on a local, continental, and intercontinental scale. Virus transport by birds can, in theory, explain the colonization of new geographic regions by arboviruses, why some arboviruses in widely separated areas are genetically similar, and how arboviruses annually recur in temperate latitudes following interrupted transmission during the winter months. The four scenarios in which a bird could transport an arbovirus include (1) a viremic bird moving while it maintains a viremia sufficient to infect an arthropod that feeds on it at a new locale; (2) a bird previously infected by an arbovirus maintaining a chronic, low-level virus infection that, perhaps because of the stresses associated with annual movement, recrudesces to produce a viremia high enough to infect an arthropod at a new locale or at a different time of year; (3) an infected bird moving and then directly transmitting the virus to other animals either by being preyed upon or scavenged or when other birds contact its saliva or feces; and (4) a bird transporting virus-infected arthropods that drop off at a new location. The idea that birds spread arboviruses is based largely on records of virus-positive birds of unknown movement status caught during the migration season, serological data showing that migrant birds were exposed to virus in the past, and indirect inferences about arbovirus movement based on patterns of genetic variation in viruses in different geographic locations. We review the direct and indirect evidence for these scenarios. Although there are a few records of migrant birds having moved arboviruses over long distances, we conclude that there is no strong empirical evidence that wild birds play a major role in the dispersal of these pathogens at the continental or intercontinental levels or that arboviruses routinely become established at new foci or are seasonally reintroduced into established foci as a result of transport by birds. Additional field and laboratory studies on how virus infection directly affects a bird's likelihood of moving are needed. Researchers interested in virus transport should focus on the extent to which birds move viruses locally and how local transport contributes to arbovirus dispersal more generally, whether virus-infected arthropod vectors disperse long distances, and the extent to which arboviruses are maintained at established foci through vertical transmission and overwintering by adult vectors. Unjustified assumptions that wild birds disperse pathogens could negatively affect the conservation of many migratory species throughout the world and cause public health resources to be diverted into ineffective ways to predict or prevent disease spread. *Received 7 May 2009, accepted 27 August 2010.*

[1]E-mail: charles-brown@utulsa.edu

[2]Present address: Department of Entomology and Plant Pathology, Oklahoma State University, Stillwater, Oklahoma 74078, USA.

Ornithological Monographs, Number 71, pages 1–64. ISBN: 978-0-943610-90-0. © 2011 by The American Ornithologists' Union.
All rights reserved. Please direct all requests for permission to photocopy or reproduce article content through the University of California Press's Rights and Permissions website, http://www.ucpressjournals.com/reprintInfo.asp. DOI: 10.1525/om.2011.71.1.1.

¿Son las Aves Silvestres Importantes para el Transporte de Virus Transmitidos por Artrópodos?

Resumen.—Los virus encefalíticos transmitidos por artrópodos (arbovirus) pueden causar una variedad de enfermedades serias en humanos y en especies silvestres, incluyendo la encefalomielitis equina occidental, la encefalitis de St. Louis, la encefalitis japonesa y la enfermedad neuroinvasora del oeste del Nilo. Entender cómo se dispersan estos patógenos en el ambiente es importante tanto para manejar su impacto en relación con la salud, como para interpretar sus patrones de variabilidad genética en áreas amplias. Debido a que muchos arbovirus infectan aves silvestres y pueden amplificarse a un nivel que hace que las aves puedan infectar a los insectos vectores, varios investigadores han sugerido que los movimientos de las aves migratorias representan una de las principales formas de transporte de estos virus a escalas locales, continentales e intercontinentales. En teoría, el transporte por las aves podría explicar la colonización de nuevas regiones geográficas por parte de los arbovirus, por qué algunos arbovirus en áreas distantes son genéticamente similares y cómo los arbovirus reaparecen anualmente en latitudes templadas luego de la interrupción de su transmisión que sucede durante los meses de invierno. Los cuatro escenarios en los que un ave podría transportar un arbovirus incluyen (1) un ave virémica que se mueve mientras mantiene una viremia suficiente para infectar a un artrópodo que se alimenta de ella en una localidad nueva; (2) un ave previamente infectada por un arbovirus que mantiene una infección viral crónica y de bajo nivel que, quizás debido al estrés asociado con los movimientos anuales, se recrudece para producir una viremia suficientemente alta para infectar un artrópodo en una localidad nueva o en otra época del año; (3) un ave infectada que se mueve y luego transmite el virus directamente a otros animales que se alimentan de ella (depredadores, carroñeros) o mediante el contacto con su saliva o heces; y (4) un ave que transporta vectores infectados con virus a un nuevo sitio. La idea de que las aves diseminan los arbovirus se basa en gran medida en registros de aves virus-positivas de estatus de movimiento desconocido capturadas durante la temporada de migración, en datos serológicos que demuestran que algunas aves migrantes estuvieron expuestas a virus en el pasado y en inferencias indirectas sobre el movimiento de los arbovirus basadas en los patrones de variación genética observados en virus de diferentes lugares geográficos. En este trabajo revisamos la evidencia directa e indirecta que apoya estos escenarios. A pesar de que existen unos pocos registros que demuestran que algunas aves migrantes han transportado arbovirus a lo largo de grandes distancias, concluimos que no existe evidencia empírica fuerte de que las aves silvestres desempeñen un papel importante en la dispersión de estos patógenos a niveles continentales o intercontinentales. Tampoco hay evidencia sólida de que los arbovirus rutinariamente se establezcan en nuevos focos ni que se reintroduzcan estacionalmente en focos establecidos como resultado del transporte por medio de las aves. Se necesitan más estudios de campo y de laboratorio acerca de cómo las infecciones virales afectan directamente la probabilidad de que las aves se muevan. Los investigadores interesados en el transporte de los virus deben enfocarse en (1) conocer el grado al que las aves mueven los virus localmente y en cómo el transporte local contribuye a la dispersión de los arbovirus de modo más general, y (2) comprender en qué grado los arbovirus se mantienen en focos establecidos gracias a la transmisión vertical y a la permanencia durante el invierno de vectores adultos. La suposición injustificada de que las aves dispersan los patógenos podría afectar negativamente la conservación de muchas especies migratorias a nivel mundial y causar que recursos para la salud pública sean desviados hacia modos no efectivos para predecir o prevenir la diseminación de enfermedades.

INTRODUCTION

When the legend becomes fact, print the legend.
—"The Man Who Shot Liberty Valance," 1962

BIRDS ARE THE primary vertebrate amplifying hosts for a number of arthropod-borne viruses (arboviruses). Upon being bitten by an infected arthropod (most often mosquitoes), some bird species replicate the virus to an extent that they can, in turn, infect transmission-competent arthropod vectors that feed on the viremic birds (Yuill 1986; Morris 1988; Scott 1988; Reisen and Monath 1989; Weaver et al. 1992, 2004; Calisher 1994; Gould et al. 2001; Weaver and Barrett 2004; McLean 2006; Unnasch et al. 2006a, b; Kramer et al. 2008). Consequently, evidence of virus amplification in birds is used to determine local virus transmission and can sometimes predict impending epidemics or epizootics (e.g., Kissling et al. 1955; Lord et al. 1974; McLean and Bowen 1980; McLean et al. 1993; Reisen et al. 2000b, 2006a; Day 2001; Eidson et al. 2001; Julian et al. 2002; Howard et al. 2004; Komar et al. 2005; Kwan et al. 2010).

By moving while viremic and being fed upon by uninfected vectors at their destinations, birds are also widely assumed to be responsible for introducing arboviruses into new geographic areas and annually reintroducing them to places where cold winter temperatures interrupt virus transmission for part of the year (Miles and Howes 1953; Johnson 1960; Stamm and Newman 1963; Lord and Calisher 1970; Calisher et al. 1971; Hannoun et al. 1972; Work and Lord 1972; Morris et al. 1973; Hayes 1989; Nice 1994; Hanna et al. 1996; Kramer et al. 1997, 2008; Rappole et al. 2000; Malkinson and Banet 2002; Peterson et al. 2003; Reed et al. 2003; Bengis et al. 2004; Gould et al. 2004; Hubálek 2004; Lvov et al. 2004; Nga et al. 2004; McLean 2006; Owen et al. 2006; Sullivan et al. 2006; Figuerola et al. 2007; Jourdain et al. 2007a, b; Linke et al. 2007; Georgopoulou and Tsiouris 2008). Birds are thought to be responsible for most cases of arbovirus transport largely because (1) it is assumed that arthropod vectors are generally relatively sedentary and do not have the long-distance movement potential of migratory birds (Brust 1980, Lillie et al. 1981, LaSalle and Dakin 1982, Milby et al. 1983, Howard et al. 1989, Fairley et al. 2000, Merrill et al. 2005, Maciel-de-Freitas et al. 2006); (2) it is assumed that overwintering of viruses in arthropod vectors or their eggs occurs infrequently in the colder, temperate latitudes (Reeves 1974, 1990; Rosen 1987; Reisen 1990; Reisen et al. 2006b; Brown et al. 2009a); and (3) observations show that some migratory birds can move over long distances very quickly (Johnson et al. 2004, Gill et al. 2005, Jourdain et al. 2007a, Stutchbury et al. 2009). That birds transport viruses over various spatial scales has almost become dogma in arbovirology and public health, and links between animal migration and the dispersal of pathogens more generally have been hypothesized (Altizer et al. 2011). However, there have been few direct observations of birds infected with arboviruses definitively undergoing long-distance movement. Most of what we think we know about transport of these viruses is based on records of viremic birds of unknown status caught during the migration season or indirect inferences about virus transport based on patterns of genetic variation of arboviruses in different geographic locations.

Although some workers have suggested that the evidence is weak for widespread arbovirus transport by birds (Takahashi et al. 1972; Emord and Morris 1984; Morris 1988; Scott 1988; Reisen et al. 2000b, 2003c, 2010; Komar and Clark 2006; Altizer et al. 2011), birds remain implicated as critical players in the transmission and dispersal of zoonotic viruses in general, as illustrated most recently by the extensive field and experimental work done on bird species competence for West Nile virus (WNV) in North America (Komar 2003, McLean 2006, Kilpatrick et al. 2007, Kramer et al. 2008, Dusek et al. 2009, Wheeler et al. 2009) and the current attention given to migratory birds as dispersers of avian influenza world-wide (e.g., Tracey et al. 2004, Gilbert et al. 2006, Kilpatrick et al. 2006a, Olsen et al. 2006, Gauthier-Clerc et al. 2007, Krauss et al. 2007, Pearce et al. 2009, Lebarbenchon et al. 2010). Much of the work on WNV in particular has assumed, without direct evidence, that the virus is introduced or reintroduced to given locales by migrating birds (Rappole et al. 2000, Petersen and Roehrig 2001, Peterson et al. 2003, Hubálek 2004, Lewis et al. 2006, McLean 2006, Kramer et al. 2008, Dusek et al. 2009), and only recently have alternatives (e.g., transport by mosquito vectors) been seriously considered (Goldberg et al. 2010, Venkatesan and Rasgon 2010).

Understanding how arboviruses are moved is important, given the medical significance of many of them. Of the 534 described arboviruses world-wide, 134 (25%) are known to cause illness in humans, and there has been a global resurgence in zoonotic arbovirus diseases in the past 25 years (Gubler 2002). Some of these diseases may be expected to increase in frequency or severity in the future with global climate change (Shope 1991, Patz and Reisen 2001, Zell 2004, Gould and Higgs 2009, Weaver and Reisen 2010). Specifying the extent to which birds transport arboviruses can assist in developing potential virus control and management strategies, help determine how public health resources can best be allocated, and prevent unwarranted assumptions that influence public perception of birds or other wildlife (Yasue et al. 2006, Weber and Stilianakis 2007).

THE MAJOR BIRD-ASSOCIATED ARBOVIRUSES

The arboviruses typically associated with birds that have been best studied include, in

North America, the alphaviruses (Togaviridae) eastern equine encephalomyelitis virus (EEEV), western equine encephalomyelitis virus (WEEV), Highlands J virus (HJV), and Buggy Creek/Fort Morgan virus (BCRV); and, in the Old World, Sindbis virus (SINV). These viruses can cause encephalitic disease, arthralgia, and mortality in humans, livestock, poultry, or wild birds (Scott et al. 1984, Morris 1988, Niklasson 1989, Reisen and Monath 1989, Scott and Weaver 1989, Calisher 1994, Cilnis et al. 1996, Laine et al. 2004, O'Brien et al. 2010a, Weaver and Reisen 2010). The well-studied flaviviruses (Flaviviridae) include, in North America, the native St. Louis encephalitis virus (SLEV) and the recently introduced West Nile virus (WNV); and, in the Old World, Japanese encephalitis virus (JEV). These flaviviruses are also serious threats to human health (McLean and Scott 1979, McLean and Bowen 1980, Hayes 1989, Gruwell et al. 2000, Rappole et al. 2000, Day 2001, Komar 2003, Reisen 2003, Marra et al. 2004, McLean 2006, Kramer et al. 2008, Weaver and Reisen 2010).

Because most hematophagous arthropods show distinct preferences for specific host taxa, virtually all arboviruses are transmitted by only a subset of the mosquito (or other arthropod) species that occur in a given area. In most cases of virus transmission by arthropods, the vector's alimentary tract is infected following a blood meal, virions are disseminated throughout the vector, and eventually the virus is replicated in the salivary glands, whereupon infectious saliva is injected into the host during later blood feeding (Weaver and Reisen 2010). Different vector species (sometimes even ones closely related) show widely varying degrees of competence for transmission of different arboviruses, usually related to taxon-specific (and sometimes individual-specific) differences in the dosage (host viremia level) needed to infect, the extent of various physiological barriers to replication or movement of virus to and from the salivary glands, and extrinsic constraints (e.g., ambient temperature and breeding phenology) on viral replication (Hardy et al. 1983, Hardy and Reeves 1990a, Lundström 1994, Goddard et al. 2002, Reisen et al. 2005a).

Alphaviruses

Widespread in the eastern United States and in Central and South America, EEEV causes the most severe human disease of any of the native arboviruses in North America. It was first isolated in 1933 in horses along the east coast of the United States, and birds were soon considered a possible reservoir host because of the timing and location of outbreaks, but the virus was not isolated from a wild bird until 1950. Since then, studies have shown that many passerines and various other birds can be infected by EEEV in its North American range (Morris 1988, Calisher 1994, Crans et al. 1994, Unnasch et al. 2006b). In North America, EEEV is transmitted mostly by the bird-feeding mosquito *Culiseta melanura*, while in South America the virus is transmitted by mammalophilic mosquitoes in the genus *Culex* (subgenus *Melanconion*) and amplified primarily in small mammals (Scott and Weaver 1989, Scott et al. 1994).

Western equine encephalomyelitis virus occurs in western North America and in Brazil and Argentina (Reisen and Monath 1989, Calisher 1994). The primary transmission cycle of WEEV involves the ornithophilic mosquito *Culex tarsalis*, and birds as the principal vertebrate host, although WEEV infection has also been detected in mammals, reptiles, and amphibians (Reisen and Monath 1989, Milby and Reeves 1990). It causes disease in humans and livestock, although cases have declined markedly in recent years (Forrester et al. 2008, Reisen et al. 2008).

Originally isolated in the eastern United States in 1952, HJV was once considered a WEEV variant. Later analyses identified it as a distinct species in the WEEV complex (Calisher et al. 1988, Reisen and Monath 1989), maintained in an epiornitic cycle involving small passerines and mostly *C. melanura* vectors.

Also closely related to WEEV, BCRV (and its variants, Fort Morgan virus and Stone Lakes virus; Pfeffer et al. 2006, Padhi et al. 2008, Brault et al. 2009) is an ecologically unusual arbovirus that is confined to Cliff Swallow (*Petrochelidon pyrrhonota*) colonies, where it is maintained by its vector, the ectoparasitic Swallow Bug (*Oeciacus vicarius*), and is amplified in Cliff Swallows and invasive House Sparrows (*Passer domesticus*) that occupy swallow nests (Hayes et al. 1977; Scott et al. 1984; Hopla et al. 1993; Brown et al. 2001, 2007, 2008, 2009c, 2010b; O'Brien et al. 2010a, 2011).

First isolated from mosquitoes collected near the village of Sindbis in Egypt in 1952, SINV is widely distributed throughout Europe, Asia, Africa, and Australia (Taylor et al. 1955, Niklasson 1989, Lundström 1999, Laine et al. 2004). It is transmitted primarily by *Culex* spp. (Lundström 1994) and has been found in various taxonomic groups of birds as well as some amphibians and mammals (Lundström et al. 1993, 2001; Lundström 1999; Hubálek 2008; Kurkela et al. 2008).

Flaviviruses

The most medically important flavivirus in temperate North America until the recent arrival of WNV, SLEV is found from Argentina to Canada, but it seems to be more common in North America, with periodic human disease outbreaks attributed to it in the United States (Monath 1980, Day 2001). It is largely transmitted by *Culex* spp. and amplified in multiple wild bird species.

West Nile virus occurs endemically in much of Africa (where it was discovered) but also circulates in Europe, Asia, and Australia (Hayes 1989, Weaver and Reisen 2010). It has been reported to infect mammals and reptiles, but birds are considered the primary amplifying hosts (McLean 2006, Kilpatrick et al. 2007, Kramer et al. 2008) and *Culex* spp. the primary vectors. In North America, WNV is an invasive arbovirus that was first detected in 1999 in New York City. The virus rapidly spread east to west across the United States (McLean 2006, Rappole et al. 2006) and then south into Central and South America (Kramer et al. 2008).

Among the more significant arboviruses worldwide is JEV, found commonly throughout Asia and responsible for more human cases annually than all other bird-associated arboviruses (Burke and Leake 1988, Weaver and Reisen 2010). It is transmitted by *Culex* spp., the most common vertebrate hosts being wild birds of multiple species, especially Ciconiiformes, and domestic pigs (Erlanger et al. 2009).

Bird Transport of Viruses: Scenarios and Consequences

There are four primary ways that a bird could be responsible for transporting an arbovirus from place to place. The most commonly discussed scenario is one in which a virus-infected bird moves and initiates transmission at a new locale by being fed upon by a hematophagous arthropod. In this case, (1) an infected arthropod vector bites a bird and transfers virus to that bird in sufficient quantity to initiate a viremia in the bird; (2) that bird then begins or continues its movement while infected, such that (3) it travels to a different location and maintains a viremia high enough to infect an arthropod that feeds on it at the new locale. All of these conditions must be met for arbovirus transport to occur via this mechanism.

In a second scenario, a bird that was previously infected by an arbovirus sometime earlier in its life maintains a chronic, low-level infection of that virus. Perhaps because of the stresses associated with annual migration, movement, or reproduction, the resultant suppression of the immune system allows the virus to recrudesce during or after the time that a bird moves, resulting in a viremia high enough to infect an arthropod that bites the bird in a new locale or at a different time of year.

A third scenario is one in which infected birds infect others through direct transmission of viruses. An infected bird could move and subsequently be preyed upon or scavenged at its new location and thus infect predators or scavengers orally via consumption of the virus-infected carcass. This could occur if the bird was maintaining a viremia at the time of its death, or if chronic persistence of noncirculating virus in tissues such as the kidney, spleen, or brain is sufficient to infect the predator or scavenger. Direct transmission could also occur through shedding of virus in saliva or feces, most likely in colonial or communally roosting birds where many individuals are concentrated in close physical contact.

The fourth scenario in which arboviruses might be moved is when virus-infected arthropods (such as ticks) are transported while attached to a bird and then drop off the bird at a new location. Direct transmission and carrying of infected arthropods are not as commonly invoked to explain arbovirus transport, especially for the mosquito-borne viruses, and fewer data are available to evaluate them.

Bird-mediated transport of arboviruses potentially has three major ecological or evolutionary consequences: (1) virus may be introduced to new geographic locales, resulting in virus colonization and potential invasion of novel hosts and environments; (2) endemic arboviruses may be interchanged over wide areas, resulting in mixing of haplotypes and genetic homogeneity of viruses from different geographic locales; and (3) endemic viruses may be seasonally reintroduced into an area where cold winter temperatures interrupt transmission by arthropods for at least part of the year. Although there are perhaps subtle behavioral or ecological differences in how birds might transport arboviruses in each of these contexts, overall the same general kinds of evidence to support bird-associated transport are needed regardless of ecological or evolutionary outcome.

Bird movement occurs on a continuum, from localized travel in and around a nesting territory or colony, to short-range movements among territories or between foraging grounds, to long-distance migratory movement across continents or between continents. Because most attention has been given to the role of birds in transporting viruses at the continental or intercontinental scales, in this review we focus primarily on the likelihood of long-distance arbovirus transport by birds known to be migratory or nomadic over distances of at least several hundred kilometers. However, we also discuss the evidence for more localized transport of arboviruses by resident birds, although we recognize that distinguishing local from "long"-distance movement can be problematic for some species and that this is a somewhat arbitrary distinction.

Scenario 1: Viremic Birds Move and Infect Vectors

Vectors Transfer Virus to Migratory Birds and Initiate Viremia

For long-range transport of arboviruses by birds to occur, migratory or nomadic species must amplify the virus well enough for transmission-competent vectors to become infected by feeding on them, and these migrant birds must become infected at an appropriate time of the year. Generalizing about host competence from studies on nonmigratory species, even ones closely related to migratory taxa, may not be appropriate. Some evidence indicates that migratory birds invest more in immune function than do closely related resident species (Møller and Erritzøe 1998), yet intense activity such as that occurring during migration can also suppress the avian immune system (Weber and Stilianakis 2007, Altizer et al. 2011). These immunological differences might directly affect the likelihood that a species will become infected by a virus upon being bitten by a vector and that it will develop a high viremia. Thus, migratory birds could be either more or less likely than resident species to amplify arboviruses and consequently might be either better or worse candidates for transporting virus.

Suitability of migrant versus resident birds as hosts: Experimental infections.—Although a few studies have focused on migratory species, such as that of Dickerman et al. (1980) on Venezuelan equine encephalitis virus (VEEV, which is not a virus typically associated with migratory birds), most of the bird species that have been studied experimentally as amplifying hosts for arboviruses are not highly migratory or nomadic. For example, Kissling et al.'s (1957) experimental infections with HJV in Louisiana used four permanent resident species and two species that are winter residents there, and early research on WNV used nonmigratory resident birds (Work et al. 1955). Much of the experimental work on WEEV, SLEV, and WNV in North America has been done with a relatively few species, such as House Sparrows, House Finches (*Carpodacus mexicanus*), White-crowned Sparrows (*Zonotrichia leucophrys*), European Starlings (*Sturnus vulgaris*), and American Robins (*Turdus migratorius*) (Hammon et al. 1951; McLean and Scott 1979; McLean et al. 1983; Hardy and Reeves 1990b; Komar et al. 1999; Reisen et al. 2000a, b, 2001, 2003a, b, 2004a, b, 2006a, b). Most of these species do not undertake regular long-distance movements, with adults (and often juveniles) typically settling within 8–15 km of their previous nesting site and some dispersing no farther than 90 m or (in the case of robins and starlings) migrating well before or largely after the peak times of transmission-competent mosquito activity (Lowther and Cink 1992, Cabe 1993, Hill 1993, Chilton et al. 1995, Sallabanks and James 1999, Rappole and Hubálek 2003). Although we know much about how they respond to arbovirus infection and how long they are potentially infectious to arthropod vectors, these particular bird species are not likely to transport arboviruses over long distances or serve as useful surrogates for migratory species.

Reisen et al. (2003c) experimentally inoculated birds of 27 species from California with WEEV and SLEV. All but six were largely resident species in the state (based on Small 1974). The results showed that the migratory species, consisting of three warbler species, two sparrow species, and the Bullock's Oriole (*Icterus bullockii*), were as likely to develop viremias as the resident species (Reisen et al. 2003c). Further work showed that two species of migratory warblers developed short-lived viremias to SLEV comparable to that of many resident species (Reisen et al. 2010). Similar results were found for Eurasian SINV (Lundström et al. 1993), in which the migratory

Common Goldeneye (*Bucephala clangula*) and at least five migratory passerine species developed viremias of sufficient length and strength to potentially infect vectors that might feed on the birds after they moved.

In a study on 25 bird species representing 17 families and 10 orders, experimental infection with WNV revealed 16 species that were deemed competent amplifying hosts, as defined by exhibiting a virus titer of at least $5.0 \log_{10}$ PFU mL^{-1} on one or more days after infection (N. Komar et al. 2003). Seven of the 16 competent host species are partially migratory in at least part of their range, although none was a Neotropical migrant that regularly migrates beyond North America. Other experimental infections with WNV (Wheeler et al. 2009, Reisen et al. 2010) showed that three migratory species, the Orange-crowned Warbler (*Vermivora celata*), the Yellow Warbler (*Dendroica petechia*), and the Common Yellowthroat (*Geothlypis trichas*), were competent hosts with peak viremia titers exceeding $6.0 \log_{10}$ PFU mL^{-1}. Most of the Orange-crowned Warblers succumbed to infection, whereas few of the Yellow Warblers did so.

Suitability of migrant versus resident birds as hosts: Field serosurveys.—Because of the low frequency with which arboviruses are isolated from individuals in most wild bird populations (Table 1), for over 50 years researchers have used the presence of antibodies to these viruses as a proxy to infer which host species are likely exposed and which ones may be important in transmission cycles. The literature on seroprevalence in different bird species is vast, and here we focus on serological work that has provided explicit comparisons of migratory and resident bird species.

Two studies on WNV in Europe found higher seroprevalence to WNV in migrant bird species than in residents (Jourdain et al. 2008, López et al. 2008), whereas in Senegal resident birds had higher levels of WNV antibodies (Chevalier et al. 2009). For WEEV and SLEV in California and SLEV in Florida, resident bird species tended to show greater field seroprevalence to these viruses than did migratory species (Day and Stark 1999, Reisen et al. 2003c). Markedly higher seroprevalence in residents than in migrants was also found for WNV in California (Wheeler et al. 2009, Reisen et al. 2010), and comparisons of migratory versus resident populations within raptor species showed a similar result (Hull et al. 2006). These studies seem to indicate, overall, a greater degree of natural virus infection in resident birds than in migratory ones.

Conclusions about host exposure or competence from serological data alone are difficult, however, because these data do not provide information on past virus titer and duration of viremia (i.e., whether the bird was ever infectious to vectors). In some species, seroconversion may occur even when some individuals do not develop detectable viremia in blood (Kissling et al. 1957; Reisen et al. 2000a, 2003c; Huyvaert et al. 2008), and thus the presence of antibodies to arboviruses in migratory species is not sufficient, by itself, for concluding that migrants are capable of generating the virus titers necessary for transmission. The problems in interpreting serological data are illustrated by a massive serosurvey of >26,000 individuals of 157 bird species for WEEV and SLEV in California (Reisen et al. 2003c). In that study, most of the apparent migrant species were placed into the never-infected category. Yet we do not know whether the absence or reduced prevalence of antibodies in migrants means that migrants (1) were never infected, (2) were infected but most died (leaving few remaining to show antibodies), or (3) degraded their antibodies more rapidly than resident species. Similar limitations of interpretation apply to a recent field serosurvey of >13,000 North American birds for WNV designed specifically to study the transport of virus (Dusek et al. 2009).

Clearly, there are differences among bird species in their competence as arbovirus hosts and the extent to which different species are fed upon by vectors of varying transmission competence; bird species thus differ widely in their potential contribution to virus transmission (McLean and Scott 1979; Dickerman et al. 1980; Reisen et al. 2000b, 2003c, 2005a; Hassan et al. 2003; Cupp et al. 2004; Kilpatrick et al. 2006b, c). On balance, the evidence indicates that some migratory species can develop high enough viremia to infect arthropod vectors, although resident bird species seem to be more commonly infected with arboviruses.

Exposure of migrants to arboviruses prior to movement.—The scenario of bird-mediated arbovirus transport requires that migrant birds be exposed to infected vectors at sites where virus occurs at least enzootically among the arthropod vectors. However, it should be emphasized that only transmission-competent taxa are relevant for either initial infection of a migrant bird or subsequent transmission to another vector.

TABLE 1. Overall prevalence of arboviruses detected in blood of wild birds (viremic individuals) in field surveys. Virus detections in other tissues were excluded from the percentages where possible.

Number of species [a]	Ages [b]	Sample size [c]	Viremic (%)	Target viruses [d]	Locality	Reference
104	Adults	1,421	0.35	EEEV, HJV	Louisiana	Kissling et al. 1955
~41	Adults	748	0.00	EEEV	Massachusetts	Hayes et al. 1962
35	Adults	649	8.6	EEEV, HJV	Alabama	Stamm and Newman 1963
—	Adults	263	6.5	EEEV, HJV	Maryland	Lord and Calisher 1970
—	Adults	118	2.5	EEEV, HJV	Virginia	Lord and Calisher 1970
—	Adults	561	0.89	EEEV, HJV	North Carolina	Lord and Calisher 1970
—	Adults	253	0.39	EEEV, HJV	Georgia	Lord and Calisher 1970
—	Adults	311	0.32	EEEV, HJV	Florida	Lord and Calisher 1970
~39	Adults	2,566	0.08	EEEV	Florida	Bond et al. 1972
~29	Adults	2,866	0.28	EEEV, HJV	Maryland	Dalrymple et al. 1972
52	Adults	467	0.00	EEEV	Texas	Work and Lord 1972
~34	Adults	1,558	0.06	EEEV	Belize	Work and Lord 1972
78	Adults	1,821	0.11	EEEV	Louisiana	Work and Lord 1972
30	Adults	212	1.4	EEEV	New York	Bast et al. 1973
5	Adults	14	0.00	EEEV	New York	Morris et al. 1973
~48	Adults	1,848	1.0	EEEV	New Jersey	Crans et al. 1994
2 [e]	Adults, nestlings	457	0.22	EEEV	Florida	Garvin et al. 2004
83	Adults	4,174	0.91	EEEV, HJV	New York	Howard et al. 2004
106	Adults	885	1.4	EEEV, SLEV, WEEV	Brazil	Woodall et al. 1972
~51 families	Adults, nestlings	32,220	0.16	SLEV	Western Hemisphere	McLean and Bowen 1980 [f]
33	Adults	363	0.00	SLEV	Arkansas	McLean et al. 1993
25	Adults	663	0.75	SLEV	Florida	Day and Stark 1999
8	Nestlings	194	0.00	SLEV, WEEV	California	Reisen et al. 2000b
1 [g]	Nestlings	3,964	5.9 (WEEV) 0.38 (SLEV)	SLEV, WEEV	Texas	Holden et al. 1973
—	Adults, nestlings	4,348	0.53	SLEV, WEEV	Colorado	Cockburn et al. 1957
44	Adults, nestlings	1,800	0.55	SLEV, WEEV	California	Milby and Reeves 1990
48 families	Adults	943	0.42	SLEV, WEEV	Brazil	Shope et al. 1966
84	Adults (winter)	3,242	0.06	SLEV, WEEV	California	Reeves 1990
20	Nestlings	1,491	0.20	WEEV	Colorado	Sooter et al. 1952
24	Adults	124	0.00	SINV	Czechoslovakia	Ernek et al. 1968
~34	Adults	2,601	0.15	WNV, SINV	Israel	Nir et al. 1967
35	Adults	169	3.0	WNV, SINV	Slovakia	Ernek et al. 1977

TABLE 1. Continued.

Number of species[a]	Ages[b]	Sample size[c]	Viremic (%)	Target viruses[d]	Locality	Reference
49	Adults	1,418	0.0	WNV, SLEV	California	Reisen et al. 2010
4	Adults	169	3.0	WNV	Egypt	Taylor et al. 1956
5	Nestlings	400	2.3	WNV	California	Reisen et al. 2005b, 2009b
57	Adults	998	1.1	WNV	Illinois	Hamer et al. 2008
12	Nestlings	194	0.51	WNV	Illinois	Loss et al. 2009
1[g]	Nestlings	173	4.0	WNV	Nebraska	O'Brien et al. 2010b
4	Nestlings, adults	2,620	2.0	JEV	Japan	Buescher et al. 1959
23	Adults	112	0.00	Tick-borne viruses	Slovakia	Ernek et al. 1973
2[h]	Nestlings	698	11.3	BCRV	Colorado	Hayes et al. 1977, Scott et al. 1984
2[h]	Adults	708	0.00[i]	BCRV	Nebraska	O'Brien et al. 2011
2[h]	Nestlings	1,527	9.2[i]	BCRV	Nebraska	O'Brien et al. 2011
11	Adults	3,044	1.9	Matruh	Egypt	Berge et al. 1971
26	Adults	717	0.56	Bahig, Matruh	Italy	Balducci et al. 1973
133	Adults	13,403	0.14 (WNV) 0.16 (EEEV, HJV)	multiple	Eastern United States	Dusek et al. 2009
12	Adults	400	0.00	multiple	Tunisia	Hannoun et al. 1972
~44	Adults	401	0.00	multiple	Uganda	Kafuko 1972
—	Adults	3,300	1.6	mulitple	Egypt, Cyprus	Watson et al. 1972
Total		103,596	0.92[j]			

[a] Some studies were imprecise as to number of species sampled or only reported the families sampled.

[b] Juveniles that had fledged were considered adults.

[c] Total samples screened, usually (but not always) corresponding to total individual birds tested.

[d] Viruses that the study targeted or the principal ones found; some studies surveyed for arboviruses generally.

[e] Blue Jay, Florida Scrub-Jay.

[f] Some may have been isolations from tissues other than blood.

[g] House Sparrow.

[h] Cliff Swallow, House Sparrow.

[i] Only Vero cell isolates tabulated as positives; additional RT-PCR positives not included.

[j] Computed on the basis of total birds sampled and total birds positive across all studies; mean per study, 1.47% ($n = 51$).

Hematophagous arthropod abundance per se at a site may not always predict transmission potential to birds; for example, the total abundance of all mosquitoes may not always correlate well with abundance of specific transmission-competent species at a site (e.g., Reisen et al. 1996, Zhong et al. 2003, Andreadis et al. 2004).

It seems probable that migrant birds in the tropics can be exposed to various arboviruses prior to their spring departure for the northern nesting grounds, given that most mosquitoes (including taxa competent to transmit most arboviruses) are generally active year-round in tropical or subtropical areas (Taylor et al. 1956, Yuill 1986). However, the few studies that have reported arbovirus isolations in birds in the tropics have shown virus primarily in tropical resident species at low prevalence (Downs et al. 1957, Aitken et al. 1964, Shope et al. 1966, Woodall et al. 1972, McLean and Bowen 1980, Spence 1980), and these birds are not likely to transport virus very far. We are aware of only one isolation of a temperate-latitude arbovirus in a migrant bird in the tropics at a time that would suggest that the bird was about to migrate north (EEEV in an Orchard Oriole [*Icterus spurius*] in Belize in March; Work and Lord 1972; Table 2).

It is likely that migrant birds are frequently exposed to arboviruses in late summer before or as they depart temperate latitudes on fall migration, because *Culex* and other transmission-competent mosquitoes are often active in late summer (Madder et al. 1983, Scott and Weaver 1989, Reisen and Reeves 1990, Andreadis et al. 2004) at about the time that migratory birds are moving (Downs et al. 1959). The virus isolations in migratory and summer-resident birds in late summer (Table 2) suggest that potential migrants can be infected before they depart. Those that are southbound in fall also may routinely move through areas where virus outbreaks occur, and some might become infected en route (Stamm and Newman 1963, Lord and Calisher 1970, Dusek et al. 2009).

INFECTED BIRDS MOVE

The hypothesis that birds transport arboviruses requires that birds that have been recently infected and that are exhibiting (or about to exhibit) a viremia move over variable distances. Thus, infection with an arbovirus must not make a bird less likely to move, and the bird's movement to a new destination must take place before its viremia declines to a level that prevents transmission to a competent arthropod vector.

Effects of arbovirus infection on movement.—We know almost nothing about how arboviruses (or other viruses; Weber and Stilianakis 2007) affect movement patterns of wild birds in nature. Some viruses can make birds visibly ill (McLean et al. 1985; Yaremych et al. 2004; Nemeth et al. 2006a, b, 2009; O'Brien et al. 2010a, b), and in these cases it seems unlikely that an ill bird is as active as or moves as far as healthy individuals. In experimental studies, birds with the highest and longest-lasting viremias, which would be the best candidates for infecting vectors during or after movement, often suffer the most morbidity and mortality when infected (Work et al. 1955; Kissling et al. 1957; Hardy and Reeves 1990b; N. Komar et al. 1999, 2003; Reisen et al. 2003b; Nemeth et al. 2006a, b; Wheeler et al. 2009). There is field evidence that infection with EEEV reduces survivorship in jays (Garvin et al. 2004), and WNV is known to drastically affect survival of crows in nature (Yaremych et al. 2004, Caffrey et al. 2005, Reisen et al. 2006a) and is suspected to affect other species in the same way (McLean 2006, LaDeau et al. 2007, Wheeler et al. 2009, O'Brien et al. 2010b). Arboviruses, however, can also produce largely asymptomatic responses in birds (Hardy and Reeves 1990b; N. Komar et al. 2003; Reisen et al. 2003c; Nemeth et al. 2006a, b; Huyvaert et al. 2008), sometimes even while a bird has high viremia (Owen et al. 2006), and in these cases normal activity and potential long-distance movements are possible.

There have been two experimental studies that examined how arbovirus infection specifically might influence movement behavior in birds. In a study with SINV, Lindström et al. (2003) found that virus-inoculated European Greenfinches (*Carduelis chloris*) exhibited less locomotive activity during the period of viremia than saline-inoculated controls (Fig. 1). The results were consistent with SINV having energetic or pathological costs to these birds (Lindström et al. 2003) and suggested that infected birds are less likely to move as far (or as often) as uninfected individuals.

Owen et al. (2006) tested migratory restlessness of caged Gray Catbirds (*Dumetella carolinensis*) and Swainson's Thrushes (*Catharus ustulatus*) that were experimentally inoculated with WNV. Both species are at least partly migratory, breeding throughout much of North America; Gray Catbirds winter from the southern United States

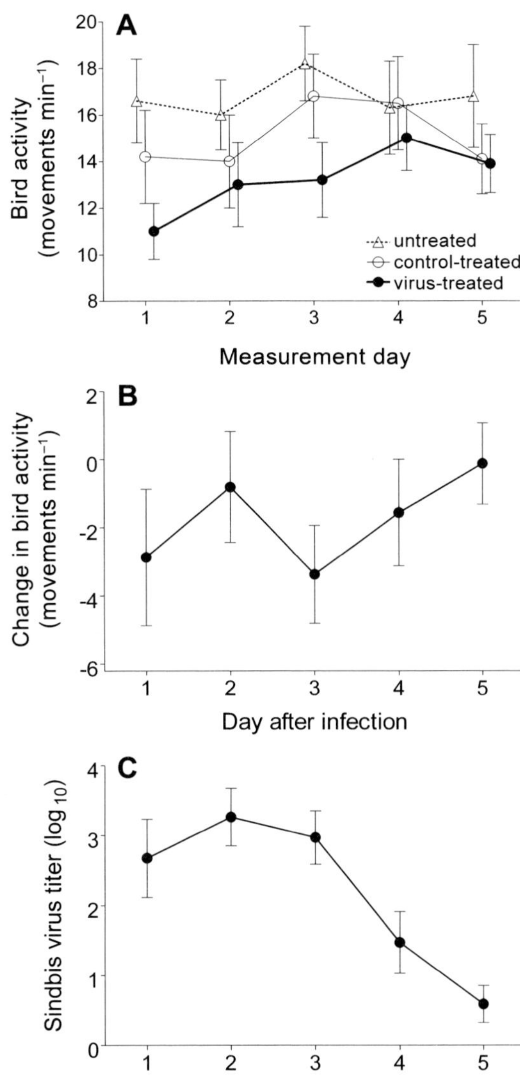

FIG. 1. (A) Bird activity, as measured from video recordings and expressed in the number of jumping or flying movements per minute, for Common Greenfinches infected with Sindbis virus, control-inoculated with saline, and not inoculated, on the days following infection; (B) the change in activity levels each day for virus-infected birds; and (C) virus titer for those infected on those days. Means ± SE are shown. (Redrawn from Lindström et al. 2003.)

south to Panama and Swainson's Thrushes from Mexico to northern South America (Cimprich and Moore 1995, Mack and Yong 2000). Viremic catbirds exhibited the same degree of autumnal migratory restlessness, as measured by activity level during times of the day when birds are generally migrating, as did control individuals (Owen et al. 2006). Migratory restlessness in both species was unrelated to viremia titers, which suggests that the birds were equally likely to migrate regardless of their level of viremia. However, in Swainson's Thrushes, some of the inoculated birds showed reduced activity levels during the viremic period (Owen et al. 2006).

These studies indicate that arboviruses can lower movement propensities of some birds, although not all viruses or bird species may behave the same. In these cases (Lindström et al. 2003, Owen et al. 2006), the birds were in captivity when tested and were given ad libitum food. Viremia might have different effects on movement behavior in nature, where the birds would need to allocate greater energy expenditure to finding food or shelter and avoiding predators.

The only work that has directly examined the effect of arbovirus infection on movement behavior of birds in the field involved American Crows (*Corvus brachyrhynchos*), a species that is a permanent resident in most areas (Yaremych et al. 2004, Ward et al. 2006). Radiotracked crows in Illinois that were naturally infected with WNV and later died seemed sluggish and did not move as far as uninfected individuals; for example, they did not commute to a nocturnal roost in the evenings, did not leave the roost to feed during the day, or did not travel as far from the roost (Yaremych et al. 2004, Ward et al. 2006). However, because crows (and other corvids) are so severely affected by WNV compared with other species, they may not represent the best model system for studying virus transport or generalizing to other birds.

An experimental field study of directly transmitted avian influenza viruses supported the presumption that infection with pathogens may reduce the movement behavior of birds. Wildcaught Bewick's Swans (*Cygnus columbianus*) in The Netherlands were experimentally infected with low-pathogenic avian influenza virus and their subsequent activities monitored with GPS collars (van Gils et al. 2007). Infected individuals had lower feeding rates, reduced fuel loads, later initiation of migration, shorter movement distances (once migration had begun), and increased use of stopover sites, compared with uninfected individuals (van Gils et al. 2007). However, a mark–recapture study of Greater White-fronted Geese (*Anser albifrons*), also in The Netherlands, found no significant difference in distance traveled between resightings during the presumed period of infectiousness for geese that tested positive versus negative for low-pathogenic avian influenza virus (Kleijn et al. 2010).

Direct field evidence for migrants carrying arboviruses.—If birds are to transport arboviruses, individuals that are migrating or otherwise moving from place to place should exhibit active viral infections. Overall, relatively few living wild birds are found with viremia, despite wide-scale sampling. Results from throughout the world reveal that, on average, <1% of the birds that have been screened for arboviruses are viremic (Table 1). Most of these studies were done in the pre-PCR era, using assay techniques thought to be less sensitive than polymerase chain reaction (e.g., Lanciotti et al. 2000, Lambert et al. 2003), although more recent work that has used RT-PCR to detect viral RNA has yielded estimates of arbovirus prevalence across species broadly similar to those of earlier research (Reisen et al. 2005b, 2009b; Hamer et al. 2008; Loss et al. 2009; O'Brien et al. 2011). Additional arbovirus isolations have been made from other tissues of birds, but because virus in organs does not necessarily lead to transmission to vectors (see below), those cases are excluded here (Table 1).

We have found reports of 999 virus detections in the blood of wild birds alive at the time of sampling and in which the migratory status of the birds tested could be ascertained with reasonable certainty (Table 2). Birds were classified as migrants if reported as such by the original authors or if the bird was found in a geographic locale where the species is predominantly neither a summer nor winter resident and the date of the isolation further suggested that the bird was in transit. If the species is known to be nonmigratory or a summer or winter resident in the locale where and when it was sampled, we classified it as a resident. If the bird was reported to be a nestling still in the nest or, in some cases, fledged but not yet capable of sustained flight, we classified it as a nestling. Most of the reported virus isolations worldwide (86.5%) have been from resident or nestling birds (Table 2). These individuals were not likely to have transported virus very far.

Sampling biases might affect the conclusion that the preponderance of virus is isolated from nonmigratory bird species (Table 2). For example, if resident birds are easier to catch and, therefore, more likely to be sampled than migrants,

we might find more isolations in resident birds even if migratory species were more likely to be infected. This is difficult to evaluate in most published studies, simply because individuals found to be virus negative are often not reported as precisely (e.g., less information on date or sampling location) as those positive for virus. For example, in the recent survey for WNV in >13,000 North American birds in the eastern United States, the sampling dates and locations were given only for the 19 individuals that were virus-positive, making it impossible to determine the percentage of migratory versus resident species that were infected at any given locale (Dusek et al. 2009).

One study that reported the data such that all birds sampled could be categorized as likely migratory or resident provides no evidence that virus detection was affected by sampling biases related to migratory status (Stamm and Newman 1963). The two arboviruses that are most commonly associated with migrant birds in North America are EEEV and HJV; in both cases, ~20% of the total isolations of these viruses were from birds that could have been migrating at the time (Table 2). Stamm and Newman (1963) sampled 649 birds in southern Alabama in late September at a time when migrants were moving through the study area and summer residents were about to depart: EEEV was isolated in 7.0% of migrants and 4.4% of summer residents, compared with 4.0% of permanent residents; HJV was found in 3.3% of migrants and 3.7% of summer residents, compared with 2.0% in permanent residents. There were no statistically significant differences between these classes of species (C. R. Brown and V. A. O'Brien unpubl. data).

Although resident birds in general may be more likely to amplify arboviruses (Table 2), evidence that migrating birds in fall could potentially transport virus southward comes from Stamm and Newman's (1963) September isolations of EEEV and HJV from transient species in Alabama. All of the migrants (mostly thrushes and warblers) and summer residents in that study (Table 2) were species that winter south of the Gulf of Mexico. Stamm and Newman (1963) speculated that these birds migrated across the Gulf of Mexico and thus potentially transported virus across the gulf, although fewer migrants appear to cross the Gulf of Mexico in fall than in spring (Able 1972), birds in general being more likely to travel along the Mexican coastline in autumn. Greater land-associated migration in fall

TABLE 2. Bird species for which arboviruses were detected from blood (potentially viremic individuals) in the field: number of individuals positive, time of sampling, locale, probable status when sampled (as reported by original authors or determined by us on the basis of species distribution and timing), and reference. Birds sampled as adults or juveniles were designated as migrant or resident (in some cases as noted, both resident and migratory populations of a given species occurred at the locale at the time of sampling, making it impossible to assign the bird definitively to either). Virus detections were by plaque assay or inoculation intracerebrally in mice or chicken eggs unless by RT-PCR as noted.

Species	Virus					
	EEEV	HJV	SLEV	WNV	WEEV	Other
PHASIANIDAE						
Greater Prairie-Chicken (*Tympanuchus cupido*)					1; August; North Dakota; **resident**; [22].	
ARDEIDAE						
Great Egret (*Ardea alba*)			1; September; Mexico; **nestling**; [29].			
Intermediate Egret (*Mesophoyx intermedia*)						JEV: 16; July–September; Japan; **nestling**; [6].
Little Egret (*Egretta garzetta*)						JEV: 3; July–September; Japan; **nestling**; [6].
Cattle Egret (*Bubulcus ibis*)				1; June; California; **nestling**; RT-PCR; [35].		
Green Heron (*Butorides virescens*)			1; September; Haiti; **resident**; [29].			
Black-crowned Night-Heron (*Nycticorax nycticorax*)				8; summer; California; **nestling**; RT-PCR; [36].		JEV: 35; July–September; Japan; **nestling**; [6].
Yellow-crowned Night-Heron (*Nyctanassa violacea*)	1; May; Louisiana; **resident**; [41].					
CHARADRIIDAE						
Northern Lapwing (*Vanellus vanellus*)						SINV: 1; March–May; Slovakia; **migrant**; [14]. Tick-borne flavivirus: 1; March–May; Slovakia; **migrant**; [14].

TABLE 2. Continued.

Species	Virus					
	EEEV	HJV	SLEV	WNV	WEEV	Other
SCOLOPACIDAE						
Green Sandpiper (*Tringa ochropus*)				1; March–May; Slovakia; **migrant**; [14].		
Lesser Yellowlegs (*T. flavipes*)	1; August; Massachusetts; **migrant**; [41].				1; May–September; Colorado; **migrant**; [9].	
LARIDAE						
Black-headed Gull (*Chroicocephalus ridibundus*)				1; March–May; Slovakia; **migrant**; [14].		
COLUMBIDAE						
Rock Pigeon (*Columba livia*)	2; September; Massachusetts; **resident**; [41].		1; August; Texas; **resident**; [29]. 1; August; Colorado; **resident**; [29]. 1; October; Florida; **resident**; [29].	2; July–August; Egypt **resident**; [43].	5; July–August; Colorado; **resident**; [9]. 7; May–September; Colorado; **nestling**; [9].	
European Turtle-Dove (*Streptopelia turtur*)				3; September; Israel; **resident**; [31].		SINV: 1; September; Israel; **resident**; [31]. Quaramfil-group flavivirus: 1; spring; Cyprus; **migrant/resident**; [44].
White-winged Dove (*Zenaida asiatica*)			1; August; Texas; **resident**; [29].			

TABLE 2. Continued.

Species	Virus					
	EEEV	HJV	SLEV	WNV	WEEV	Other
Mourning Dove (*Zenaida macroura*)			2; August; Florida; **resident**; [12]. 1; October; California; **migrant/resident**; [29]. 1; August; California; **resident**; [30].		1; August; California; **resident**; [30].	
White-tipped Dove (*Leptotila verreauxi*)			1; August; Trinidad; **nestling**; [29].			
CUCULIDAE Yellow-billed Cuckoo (*Coccyzus americanus*)	1; May; New Jersey; **resident**; [10].					
CAPRIMULGIDAE Eastern Whip-poor-will (*Caprimulgus vociferus*)	1; September; Maryland; **migrant**; [27].					
PICIDAE Downy Woodpecker (*Picoides pubescens*)	1; September; Maryland; **resident**; [11].	1; May–September; New York; **resident**; [21].		1; September; Illinois; **resident**; [13].		
Northern Flicker (*Colaptes auratus*)	3; September; New York; **migrant/resident**; [2].		1; October; Kentucky; **nestling**; [29].	1; August; Illinois; **resident**; RT-PCR; [16].		
FURNARIIDAE Olive-backed Foliage-gleaner (*Automolus infuscatus*)			1; June; Brazil; **resident**; [39].			

TABLE 2. Continued.

Species	Virus					
	EEEV	HJV	SLEV	WNV	WEEV	Other
THAMNOPHILIDAE						
Cinereous Antshrike (*Thamnomanes caesius*)			1; July; Brazil; **resident**; [39].			
Amazonian Antshrike (*Thamnophilus amazonicus*)						Turlock virus: 1; not stated; Brazil; **resident**; [46].
White-shouldered Antshrike (*Thamnophilus aethiops*)					1; not stated; Brazil; **resident**; [46].	Itaporanga virus: 1; not stated; Brazil; **resident**; [46].
Plain-throated Antwren (*Myrmotherula hauxwelli*)			1; June; Brazil; **resident**; [39].		1; June; Brazil; **resident**; [39].	
Gray Antwren (*M. menetriesii*)						Turlock virus: 1; not stated; Brazil; **resident**; [46].
White-shouldered Fire-eye (*Pyriglena leucoptera*)			1; not stated; Brazil; **resident**; [46].		1; not stated; Brazil; **resident**; [46].	Turlock virus: 1; not stated; Brazil; **resident**; [46].
FORMICARIIDAE						
Black-faced Antthrush (*Formicarius analis*)			1; not stated; Brazil; **resident**; [46].			
TYRANNIDAE						
Ochre-bellied Flycatcher (*Mionectes oleagineus*)			1; not stated; Brazil; **resident**; [46].			
Yellow-bellied Flycatcher (*Empidonax flaviventris*)	1; July–September; New York; **resident**; [20].					
Acadian Flycatcher (*E. virescens*)	1; September; Alabama; **resident**; [42].	1; September; Alabama; **resident**; [42].				
Willow Flycatcher (*E. traillii*)	1; July–September; New York; **resident**; [20].					

Table 2. Continued.

Species	Virus					
	EEEV	HJV	SLEV	WNV	WEEV	Other
Least Flycatcher (*E. minimus*)	2; July–September; New York; **resident**; [20].					
Eastern Phoebe (*Sayornis phoebe*)		1; August–September; New Jersey; **resident**; [41].				
Eastern Kingbird (*Tyrannus tyrannus*)	1; August; New Jersey; **resident**; [41].					
PIPRIDAE						
Black-and-white Manakin (*Manacus manacus*)			1; August; Trinidad; **resident**; [29].			
Blue-backed Manakin (*Chiroxiphia pareola*)			1; not stated; Brazil; **resident**; [46].			
White-crowned Manakin (*Pipra pipra*)			1; not stated; Brazil; **resident**; [46].			
ORIOLIDAE						
Eurasian Golden-Oriole (*Oriolus oriolus*)						BAHV: 1; fall; Egypt; **migrant**; [44].
LANIIDAE						
Loggerhead Shrike (*Lanius ludovicianus*)		1; June; Louisiana; **resident**; [24, 41].				

TABLE 2. Continued.

Species	Virus					
	EEEV	HJV	SLEV	WNV	WEEV	Other
VIREONIDAE						
White-eyed Vireo (*Vireo griseus*)	2; July–September; Alabama; **resident**; [41, 42].	1; July; Maryland; **resident**; [11].				
Yellow-throated Vireo (*V. flavifrons*)	1; August; New Jersey; **resident**; [41].	1; August; New Jersey; **resident**; [41].				
Blue-headed Vireo (*V. solitarius*)	1; July–September; New York; **migrant/resident**; [20].					
Philadelphia Vireo (*V. philadelphicus*)	1; September; Maryland; **migrant**; [27].					
Red-eyed Vireo (*V. olivaceus*)	3; September; Maryland; **migrant/resident**; [27]. 3; July–September; New York; **resident**; [20]. 3; September; Alabama; **resident**; [42].	1; August; Virginia; **migrant/resident**; [27]. 1; October; Maryland; **migrant/resident**; [11]. 1; May–September; New York; **resident**; [21]. 3; September; Alabama; **resident**; [42]. 1; August–September; New Jersey; **resident**; [41].				

TABLE 2. Continued.

Species	Virus					
	EEEV	HJV	SLEV	WNV	WEEV	Other
CORVIDAE						
Blue Jay (*Cyanocitta cristata*)	1; August; New Jersey; **resident**; [10].	1; August; North Carolina; **resident**; [27].	1; August; Texas; **resident**; [29].			
Florida Scrub-Jay (*Aphelocoma coerulescens*)	1; May; Florida; **nestling**; [15].					
European Magpie (*Pica pica*)				3; all year; United Kingdom; **resident**; [5].		
Black-billed Magpie (*P. hudsonia*)					1; June; Colorado; **nestling**; [40]. 1; July–August; Colorado; **resident**; [9].	
American Crow (*Corvus brachyrhynchos*)		1; August–September; New Jersey; **resident**; [41].			1; July–August; Colorado; **resident**; [9].	
Carrion Crow (*C. corone*)				1; August; Egypt; **resident**; [43].		
HIRUNDINIDAE						
Cliff Swallow (*Petrochelidon pyrrhonota*)					2; July–August; Colorado; **resident**; [9].	BCRV: 19; May–August; Colorado; **nestling**; [17, 38]. BCRV: 14; June; Nebraska; **nestling**; [33].[a] BCRV: 4; June; Nebraska; **resident**; RT-PCR [33].
Barn Swallow (*Hirundo rustica*)					8; June–August; California; **nestling**; [22, 30].	BAHV: 1; spring; Egypt; **migrant**; [44].

TABLE 2. Continued.

Species	Virus					
	EEEV	HJV	SLEV	WNV	WEEV	Other
PARIDAE						
Carolina Chickadee (*Poecile carolinensis*)	4; May–August; New Jersey; **resident**; [10, 41].	1; June; Louisiana; **resident**; [41].				
Black-capped Chickadee (*P. atricapillus*)	1; May–September; New York; **resident**; [21].					
Tufted Titmouse (*Baeolophus bicolor*)	1; July; New Jersey; **resident**; [10].	1; July; Maryland; **resident**; [11]. 1; September; Alabama; **resident**; [42].		1; September; Massachusetts; **resident**; [13].		
SITTIDAE						
Red-breasted Nuthatch (*Sitta canadensis*)	1; September; Maryland; **migrant**; [27].					
TROGLODYTIDAE						
Carolina Wren (*Thryothorus ludovicianus*)	2; September; Alabama; **resident**; [42]. 1; August–September; New Jersey; **resident**; [41].	1; September; Maryland; **resident**; [27].		1; August; Louisiana; **resident**; [26].		
House Wren (*Troglodytes aedon*)	1; July–September; New York; **resident**; [20].			1; August; Illinois; **nestling**; RT-PCR; [28].		

TABLE 2. Continued.

Species	Virus					
	EEEV	HJV	SLEV	WNV	WEEV	Other
SYLVIIDAE						
Willow Warbler (*Phylloscopus trochilus*)						BAHV: 3; fall; Egypt; **migrant**; [44].
Eurasian Chiffchaff (*P. collybita*)						BAHV: 5; fall; Egypt; **migrant**; [44].
Blackcap (*Sylvia atricapilla*)						BAHV: 1; fall; Egypt; **migrant**; [44].
Garden Warbler (*S. borin*)						BAHV: 2; fall; Egypt; **migrant**; [44]. Matariya virus; 1; October; Egypt; **migrant**; [3].
Greater Whitethroat (*S. communis*)						BAHV: 11; fall; Egypt; **migrant**; [44].
Lesser Whitethroat (*S. curruca*)						BAHV: 6; fall; Egypt; **migrant**; [44]. Simbu-group bunyavirus; 4; fall; Egypt; **migrant**; [44]. Matruh virus; 1; September; Egypt; **migrant**; [3].[b] Matariya virus; 1; October; Egypt; **migrant**; [3]. Burg el arab virus; 1; October; Egypt; **migrant**; [3].
Barred Warbler (*S. nisoria*)				1; spring; Cyprus; **migrant**; [44].		
Rueppell's Warbler (*Sylvia rueppelli*)						BAHV: 1; fall; Egypt; **migrant**; [44].

TABLE 2. Continued.

Species	Virus					
	EEEV	HJV	SLEV	WNV	WEEV	Other
MUSCICAPIDAE						
Spotted Flycatcher (*Muscicapa striata*)						BAHV: 1; fall; Egypt; **migrant**; [44]. Ingwavuma virus: 2; spring; Cyprus; **migrant**; [44].
Thrush Nightingale (*Luscinia luscinia*)						BAHV: 1; fall; Egypt; **migrant**; [44].
Common Redstart (*Phoenicurus phoenicurus*)						BAHV: 4; fall; Egypt; **migrant**; [44]. Kemerovo tick-borne virus: 1; September; Egypt; **migrant**; [37].
Whinchat (*Saxicola rubetra*)						BAHV: 1; fall; Egypt; **migrant**; [44].
TURDIDAE						
Veery (*Catharus fuscescens*)	1; September; Virginia; **migrant**; [27]. 1; July–September; New York; **resident**; [20]. 6; September; Alabama; **migrant**; [42].	1; August; North Carolina; **migrant**; [27]. 2; May–September; New York; **resident**; [21]. 6; September; Alabama; **migrant**; [42].				
Gray-cheeked Thrush (*C. minimus*)	2; September; Alabama; **migrant**; [42].	2; September; Alabama; **migrant**; [42].				

TABLE 2. Continued.

Species	Virus					
	EEEV	HJV	SLEV	WNV	WEEV	Other
Swainson's Thrush (*C. ustulatus*)	1; September; Maryland; **migrant**; [27]. 1; September; Alabama; **migrant**; [42].			1; October; Louisiana; **migrant**; [13].		
Hermit Thrush (*C. guttatus*)	1; March; Louisiana; **resident**; [24, 41].					
Wood Thrush (*Hylocichla mustelina*)	1; September; Maryland; **migrant/resident**; [27]. 1; August; New Jersey; **resident**; [10]. 1; May; Louisiana; **migrant**; [7]. 7; September; Alabama; **resident**; [42].	6; September; Alabama; **resident**; [42].		1; October; Louisiana; **migrant/resident**; [13].		
Song Thrush (*Turdus philomelos*)						BAHV: 1; spring; Egypt; **migrant**; [44].
Cocoa Thrush (*T. fumigatus*)			1; August; Trinidad; **resident**; [29].			
Bare-eyed Thrush (*T. nudigenis*)			1; August; Trinidad; **resident**; [29].			
American Robin (*T. migratorius*)	2; August–September; New Jersey; **resident**; [10].					

TABLE 2. Continued.

Species	Virus					
	EEEV	HJV	SLEV	WNV	WEEV	Other
MIMIDAE						
Gray Catbird (*Dumetella carolinensis*)	1; October; Florida; **migrant/resident**; [27]. 2; June–September; New Jersey; **resident**; [10, 41]. 3; July–September; New York; **resident**; [20]. 5; September; Alabama; **resident**; [42]. 1; April; Louisiana; **resident**; [24, 41]. 1; September; Massachusetts; **migrant/resident**; [41].	1; May– September; New York; **resident**; [21].		1; September; Massachusetts; **migrant/resident**; [13]. 2; September; New York; **migrant/ resident**; [13]. 3; September; New Jersey; **migrant/ resident**; [13]. 1; October; Virginia; **migrant/resident**; [13]. 1; October; Louisiana; **migrant/resident**; [13].		
Northern Mockingbird (*Mimus polyglottos*)	1; September; Maryland; **resident**; [27]. 1; March; Louisiana; **resident**; [41].		1; August; Texas; **resident**; [29]. 1; August; Jamaica; **nestling**; [29].			
Brown Thrasher (*Toxostoma rufum*)	1; September; Alabama; **resident**; [42].			1; October; Virginia; **resident**; [13].		
STURNIDAE						
Brahminy Starling (*Temenuchus pagodarum*)						Kammavanpettai virus; 1; March; India; **resident**; [3].
European Starling (*Sturnus vulgaris*)						SINV: 1; March–May; Slovakia; **migrant**; [14].

Table 2. Continued.

Species	Virus					
	EEEV	HJV	SLEV	WNV	WEEV	Other
PARULIDAE						
Pine Warbler (*Dendroica pinus*)	1; May; New Jersey; **resident**; [10].					
Blackpoll Warbler (*D. striata*)	1; May; Louisiana; **migrant**; [7].					
Black-and-white Warbler (*Mniotilta varia*)	1; September; New Jersey; **resident**; [10]. 2; fall; Florida; **migrant**; [4].					
American Redstart (*Setophaga ruticilla*)	1; September; Maryland; **migrant**; [27]. 1; September; New Jersey; **resident**; [10]. 1; April; Florida; **migrant**; [45].			1; September; Virginia; **migrant/ resident**; [13].		
Ovenbird (*Seiurus aurocapilla*)	1; September; Maryland; **migrant**; [27]. 2; August; New Jersey; **resident**; [10]. 3; September; Alabama; **migrant**; [42].					
Northern Waterthrush (*S. noveboracensis*)	1; September; Maryland; **migrant**; [27]. 1; August–September; New Jersey; **migrant/ resident**; [41].			1; September; New Jersey; **migrant**; [13].		

TABLE 2. Continued.

Species	Virus					
	EEEV	HJV	SLEV	WNV	WEEV	Other
Kentucky Warbler (*Oporornis formosus*)	1; August; Alabama; **resident**; [41, 42].	1; September; Alabama; **resident**; [42].				
Connecticut Warbler (*O. agilis*)	1; September; Maryland; **migrant**; [27].					
Common Yellowthroat (*Geothlypis trichas*)	1; September; New Jersey; **resident**; [10]. 5; July–September; New York; **resident**; [20, 21].					
THRAUPIDAE						
Scarlet Tanager (*Piranga olivacea*)	1; September; Maryland; **resident**; [27].					
Silver-beaked Tanager (*Ramphocelus carbo*)			1; August; Trinidad; **nestling**; [29].			
EMBERIZIDAE						
Eastern Towhee (*Pipilo erythrophthalmus*)	1; October; Maryland; **resident**; [11].			1; October; Virginia; **migrant/resident**; [13].		
Field Sparrow (*Spizella pusilla*)	1; September; Maryland; **resident**; [27].					
Song Sparrow (*Melospiza melodia*)	10; July–September; New York; **resident**; [20, 21]. 1; August–September; New Jersey; **resident**; [41].	2; May–September; New York; **resident**; [21].				
Swamp Sparrow (*M. georgiana*)	3; October; Maryland; **migrant**; [11].					

TABLE 2. Continued.

Species	Virus					
	EEEV	HJV	SLEV	WNV	WEEV	Other
White-throated Sparrow (*Zonotrichia albicollis*)	1; July–September; New York; **resident**; [20].	1; May–September; New York; **resident**; [21].				
White-crowned Sparrow (*Z. leucophrys*)					2; December–January; California; **resident**; [34].	
CARDINALIDAE Northern Cardinal (*Cardinalis cardinalis*)	1; September; Virginia; **resident**; [27]. 1; September; North Carolina; **resident**; [27]. 3; July–September; Alabama; **resident**; [41, 42]. 1; July; Louisiana; **resident**; [24, 41].	1; September; Alabama; **resident**; [42]. 1; June; Louisiana; **resident**; [24, 41].	1; October; Florida; **resident**; [12].	1; September; New Jersey; **resident**; [13].		
Rose-breasted Grosbeak (*Pheucticus ludovicianus*)	1; September; Maryland; **migrant**; [27].					
Indigo Bunting (*Passerina cyanea*)				1; September; Illinois; **migrant/resident**; [13]. 1; October; Louisiana; **migrant/resident**; [13].		
Painted Bunting (*P. ciris*)	1; October; Georgia; **migrant/resident**; [27].					

TABLE 2. Continued.

Species	Virus					
	EEEV	HJV	SLEV	WNV	WEEV	Other
ICTERIDAE						
Red-winged Blackbird (*Agelaius phoeniceus*)	1; June; New Jersey; **resident**; [10].			1; August; Illinois; **resident**; RT-PCR; [16].	2; June; Colorado; **nestling**; [40].	
Yellow-headed Blackbird (*Xanthocephalus xanthocephalus*)					1; May–September; Colorado; **resident**; [9].	
Common Grackle (*Quiscalus quiscula*)	1; June; Louisiana; **resident**; [23, 41].	1; May; Louisiana; **resident**; [41].	2; August; Florida; **resident**; [12].			
Orchard Oriole (*Icterus spurius*)	1; March; Belize; **migrant/resident**; [47].					Mayaro virus: 1; April; Louisiana; **migrant**; [8].
Yellow-rumped Cacique (*Cacicus cela*)	1; not stated; Brazil; **resident**; [46].					
FRINGILLIDAE						
Common Chaffinch (*Fringilla coelebs*)						BAHV: 1; November; Italy; **migrant/resident**; [1]. Matruh; 1; November; Italy; **migrant/resident**; [1].
Brambling (*F. montifringilla*)						BAHV: 1; October; Italy; **migrant**; [1]. Matruh; 1; November; Italy; **migrant**; [1].
House Finch (*Carpodacus mexicanus*)				2; August; Illinois; **resident**; RT-PCR; [16].	1; June; California; **resident**; [30]. 1; August; California; **nestling**; [22].	Turlock virus: 1; July; California; **nestling**; [30].
European Greenfinch (*Carduelis chloris*)						BAHV: 1; fall; Egypt; **migrant**; [44].

TABLE 2. Continued.

Species	Virus					
	EEEV	HJV	SLEV	WNV	WEEV	Other
PASSERIDAE						
House Sparrow (*Passer domesticus*)	1; September; North Carolina; **resident**; [27]. 1; August; Massachusetts; **resident**; [41]. 1; August–September; New Jersey; **resident**; [41].	2; August; New Jersey; **resident**; [41]. 2; September; New Jersey; **resident**; [18].	2; September; Illinois; **resident**; [25, 29]. 3; August; Texas; **resident**; [29]. 15; August–September; Texas; **nestling**; [19]. 3; July; Mississippi; **nestling**; [29].	7; August; Illinois; **resident**; RT-PCR; [16]. 7; August; Nebraska; **nestling**; RT-PCR; [32].	18; June–August; California; **nestling**; [22]. 234; summer; Texas; **nestling**; [19]. 3; July–August; Colorado; **resident**; [9]. 2; May–September; Colorado; **nestling**; [9]. 1; June; California; **nestling**; [30]. 1; July; California; **resident**; [30].	BCRV: 60; May–August; Colorado; **nestling**; [17, 38]. BCRV: 180; May–August; Nebraska; **nestling**; [33].c BCRV: 4; May–August; Nebraska; **resident**; RT-PCR [33].
Total species	57	23	24	26	16	32
Total isolations	139	48	53	60	297	402
Migrants	29	9	0	5	1	57
Residents	96	37	29	24	22	14
Migrants/residents	13	2	1	14	0	3
Nestlings	1	0	23	17	274	328

References: [1] Balducci et al. 1973; [2] Bast et al. 1973; [3] Berge et al. 1971; [4] Bond et al. 1972; [5] Buckley et al. 2003; [6] Buescher et al. 1959; [7] Calisher et al. 1972; [8] Calisher et al. 1974; [9] Cockburn et al. 1957; [10] Crans et al. 1994; [11] Dalrymple et al. 1972; [12] Day and Stark 1999; [13] Dusek et al. 2009; [14] Emek et al. 1977; [15] Garvin et al. 2004; [16] Hamer et al. 2008; [17] Hayes et al. 1977; [18] Holden 1955a; [19] Holden et al. 1973; [20] Howard et al. 1996; [21] Howard et al. 2004; [22] Johnson 1960; [23] Kissling et al. 1951; [24] Kissling et al. 1955; [25] Kokernot et al. 1969; [26] Komar et al. 2005; [27] Lord and Calisher 1970; [28] Loss et al. 2009; [29] McLean and Bowen 1980; [30] Milby and Reeves 1990; [31] Nir et al. 1967; [32] O'Brien et al. 2010b; [33] O'Brien et al. 2011; [34] Reeves 1990; [35] Reisen et al. 2005b; [36] Reisen et al. 2009b; [37] Schmidt and Shope 1971; [38] Scott et al. 1984; [39] Shope et al. 1966; [40] Sooter et al. 1952; [41] Stamm 1958; [42] Stamm and Newman 1963; [43] Taylor et al. 1956; [44] Watson et al. 1972; [45] Wellings et al. 1972; [46] Woodall et al. 1972; and [47] Work and Lord 1972. Unpublished records for additional resident species from Brazil from which SLEV was isolated are given in McLean and Bowen (1980).

a Nine detections by RT-PCR only.

b Fifty-seven total isolations of Matruh virus from mostly migrant birds in Egypt were reported [3], probably from blood, but sampling dates and tissue used for isolations were not stated.

c Forty detections by RT-PCR only.

would likely lower the likelihood of virus transport to Central or South America, assuming that individuals move more slowly along the Central American isthmus than they do trans-Gulf.

In contrast to the relatively high percentage of migrants with EEEV in Alabama in fall (Stamm and Newman 1963; Table 1), Bond et al. (1972) isolated EEEV from only two individuals, both Black-and-white Warblers (*Mniotilta varia*), out of >2,000 fall migrants of various species collected in Florida. No viruses were found in 566 birds during spring migration at the same site. In Lord and Calisher's (1970) sampling of 1,506 birds at seven sites along the Atlantic coast (Table 1) from August to October, EEEV in total was isolated from 16 birds and HJV from two. Some of the viremic birds were vireos, thrushes, and warblers (Table 2) that may have been migrating south to Central America or the West Indies. Lord and Calisher (1970) argued that their data indicated substantial southward transport of these viruses, and hypothesized that dispersal of EEEV might occur in short bursts when migrants are infected en route and move the virus southward incrementally.

A study of EEEV and HJV in central New York State that included >4,000 blood samples from 83 species cast doubt on the importance of bird-mediated transport for these arboviruses (Howard et al. 2004). Of the 30 isolations of EEEV and 8 of HJV, only 1 was from a species that did not breed in the study area and that could have been a transient when caught, the Blue-headed (Solitary) Vireo (*Vireo solitarius*). All virus-positive birds were caught in late summer at about the time, or after, virus was first detected in local mosquitoes, and Howard et al. (2004) suggested that these birds (including the vireo) were infected locally and that southward transport of virus into their study area from farther north likely did not occur. Another study of EEEV in upstate New York documented higher prevalence of antibodies in permanent and summer resident species than in migrants or winter residents and higher antibody prevalence in site-faithful individuals that returned between years (Emord and Morris 1984). The authors concluded that the virus circulated on site well before fall migration started and that migrating birds could not have introduced it from farther north. Similar results were found in Michigan (McLean et al. 1985) and New Jersey (Crans et al. 1994), in which no EEEV or HJV was isolated in late summer from any transient species.

It appears that SLEV, WEEV, and WNV are relatively rarely associated with migratory birds (Table 2). No definitive cases of SLEV in a bird that was clearly migrating are known, and only one isolate of WEEV (of ~300 reported) was from a migrant (Table 2). Despite the large numbers of dead birds positive for WNV reported in North America (McLean 2006), there are relatively few cases of WNV isolated from blood of a live bird that was clearly migrating at the time of sampling. Of 19 birds found viremic for WNV in late summer and fall, 2002–2003, in the eastern United States (Dusek et al. 2009), only 2 could be positively classified as migrants when sampled (Table 2). The most common species represented in the North American study was the Gray Catbird (8 of the 19 isolations; 42%), which is both a resident and a transient at each sampling site. This means that the infected birds could have been of either local or nonlocal origin (Dusek et al. 2009). The same study found no birds positive for virus in spring. Three other cases of probable migrants being found with WNV were reported from Europe (Table 2). A survey of birds migrating through the Volga River delta in southern Russia, a major passage for birds en route to and from wintering areas in Africa, revealed ~10% positive for WNV RNA, as detected in brain and spleen tissue (Lvov et al. 2004), although, because blood was not sampled, the number of viremic individuals (if any) was not known.

Two poorly known bunyaviruses of the Old World, Bahig virus (BAHV; Bunyaviridae: *Orthobunyavirus*) and its close serological relative Matruh virus, appear to be associated with migrating birds relatively often and perhaps at a greater frequency than any of the better-studied arboviruses. From >3,900 birds migrating through Egypt and Cyprus, where migratory species funnel across the Mediterranean when traveling between Europe and Africa, Watson et al. (1972) had ~40 isolations of BAHV, mostly from fall migrants and mostly from the Old World warbler genus *Sylvia*. Lesser Whitethroats (*S. curruca*) had a relatively high infection prevalence of 9.3%, illustrating apparent differences among species in proclivity for potentially transporting this virus (Table 2). Almost 60 isolations of Matruh virus were made from mostly migratory birds in Egypt (Berge et al. 1971). These birds may have been infected locally prior to sampling, however, because both of these bunyaviruses are endemic in Egypt, where they can be transmitted by ticks, some of which parasitize birds (Converse et al. 1974). The fact that BAHV and Matruh virus have apparently remained localized in the Mediterranean area even though relatively large numbers of migratory birds are infected suggests that birds do not disperse these viruses very far.

The best evidence for a role of migratory birds in long-distance arbovirus transport comes from two isolations of EEEV from a Blackpoll Warbler (*Dendroica striata*) and a Wood Thrush (*Hylocichla mustelina*) in southern Louisiana in early May, 1966 (Calisher et al. 1971). The virus isolated from these two birds proved to be South American serotypes and serologically quite unlike North American EEEV strains, which seemingly ruled out local infection after the birds' arrival in Louisiana. Because these birds were apparently migrating at the time of sampling and had likely departed from the Yucatan, western Cuba, or northern Honduras on a nonstop flight across the Gulf of Mexico, they represent a probable case of EEEV movement between Central America and the U.S. mainland (Calisher et al. 1971).

Another convincing observation of an arbovirus likely being transported by a bird over long distances, again in spring, was the finding of the tropical alphavirus Mayaro virus (MAYV) in an Orchard Oriole on the Mississippi River delta in spring 1967 (Calisher et al. 1974). Occurring in Central America and northern South America south to Amazonia, MAYV is probably most often amplified in mammals but also in birds (de Thoisy et al. 2003, Weaver and Reisen 2010). It is not known to regularly occur in the United States, and Calisher et al. (1974) interpret their finding to represent an incidental introduction of this tropical virus by a bird that had likely wintered in Latin America, where MAYV is endemic.

The finding of South American EEEV strains and the isolation of MAYV clearly illustrate the potential for birds to occasionally transport viruses from tropical to temperate areas. However, neither South American EEEV nor MAYV have become established in the United States. This suggests that these viruses are probably not regularly moved into North America by migrant birds, although their failure to colonize temperate latitudes could also reflect an absence of transmission-competent vectors among native North American mosquitoes (see below).

Limitations of existing data.—The fundamental problem with most observations of viremic migrant birds is that we do not know where the birds were infected. Unless the virus isolates can be determined to be genetically distinct from local strains (Calisher et al. 1971) or they are of a virus not regularly found in a particular area (Calisher et al. 1974), the alternative explanation of the bird being infected locally cannot be ruled out (Work and Lord 1972). For example, Lord and Calisher (1970) detected EEEV and HJV in resident species and in farmyard birds at the same time that they isolated these viruses from migrants in the same area, and they noted that some of the migrants could have been infected while traveling through areas that served as transmission foci. If a bird during migration becomes infected at or near the site it was captured and then does not move on subsequent days while it is recovering from the infection, virus is not transported anywhere (Owen et al. 2006). With a few exceptions (Calisher et al. 1971, 1974), this caveat applies to almost all of the isolations of arboviruses in apparent migrants (Table 2), because in no case was it known whether the infected bird continued, or would have continued (had it not been collected), to move.

An often-cited report of birds potentially transporting arboviruses is that of Malkinson et al. (2002) on White Storks (*Ciconia ciconia*) in Israel. Normally this species does not migrate through Israel in large numbers in autumn, but in August 1998 a flock of ~1,200 migrating birds was blown eastward off course and landed in the town of Eliat. Many of the birds were exhausted, and some died. Thirteen bird carcasses were tested for WNV, and four were positive by plaque assay and nine positive by RT-PCR (Malkinson et al. 2002). The results are suggestive that the migrating birds were infected with WNV (likely from Europe) when they arrived, because they were collected within 2 days of their appearing in Israel. Some of the storks apparently recovered and continued migrating. However, these data do not conclusively demonstrate ecologically relevant arbovirus transport by birds. The only tissue tested was brain (Malkinson et al. 2002), and the virus would have to be circulating in blood at a level sufficient to infect mosquitoes en route for it to be introduced naturally by this event. Also, the birds could have become infected in Israel near the site of collection, where WNV is endemic (Komar 2003). The observation nevertheless illustrates the potential for arbovirus introduction to new areas by birds during (perhaps sometimes errant) migratory movements.

Some studies have detected antibodies in migrant birds and used the presence of seropositive individuals to suggest that birds do or do not transport arboviruses (Emord and Morris 1984, Malkinson and Banet 2002, Dupuis et al. 2003, O. Komar et al. 2003, Linke et al. 2007, Jourdain et al. 2008, López et al. 2008, Chevalier et al. 2009). For example, up to 22% of Red-winged Blackbirds (*Agelaius phoeniceus*) in North Dakota had WNV antibodies in August (Sullivan et al. 2006). Because Red-winged Blackbirds migrate southward out of North Dakota in winter, it was suggested that this species could be an important WNV transport agent (Sullivan et al. 2006). However, the fact that so many birds had antibodies in August as they were preparing to migrate suggests only that these birds had resided in a local focus of WNV amplification prior to capture, not that viremic birds were moving or about to move. Furthermore, these birds would be unlikely to carry virus in large numbers because so many individuals at that time were immune to infection. In two studies involving 25–32 species captured in France and Spain, trans-Saharan migrants had higher average antibody prevalence for WNV than resident species, and the authors suggested that the results were consistent with migrants being more exposed to virus on their African wintering grounds and thus more likely to transport virus (Jourdain et al. 2008, López et al. 2008). In all of these studies, though, without isolations of virus from birds about to move or in the process of moving, it is difficult to conclude anything about WNV transport, simply because we cannot know when or where the birds were infected or whether they ever developed titers sufficient to infect a transmission-competent mosquito.

The finding of resident birds in Jamaica, Puerto Rico, the Dominican Republic, Mexico, and Argentina with antibodies to WNV was used to suggest that these birds were infected following local introduction of WNV to the site by birds migrating from North America to wintering grounds in Latin America (Dupuis et al. 2003, O. Komar et al. 2003, Diaz et al. 2008). A few North American migrants with antibodies were also detected, but no migrant birds with actively circulating virus were found, and thus how the virus reached these sites is not known with certainty. The large literature on seroprevalence for WNV and other arboviruses in birds may reveal what species are potentially infected, but it should be emphasized that these data cannot be used to conclude

anything about whether virus is transported by individual birds prior to seroconversion. These cautions about using serological data have been voiced previously (Komar and Clark 2006) but were largely ignored; in addition, Komar and Clark emphasized that some WNV-seropositive birds in Latin America instead may have reflected cross-reactivity with unknown ("WNV-like") enzootic flaviviruses that circulate locally.

Negative evidence.—Some studies have found little to no evidence of arbovirus-positive birds either during migration or soon after arrival on the breeding grounds (Bond et al. 1972; Hannoun et al. 1972; Kafuko 1972; Reisen et al. 2000b, 2004b, 2010; see Table 1). For example, in a study designed explicitly to search for virus in birds soon after arrival on the breeding grounds, O'Brien et al. (2008) found no evidence that migratory Cliff Swallows arrived with BCRV (also see Hayes et al. 1977), an alphavirus that is adapted specifically to Cliff Swallows and their ectoparasites (Moore et al. 2007, Brown et al. 2008, Padhi et al. 2008).

More than 1,000 migratory birds passing through the Coachella Valley of southern California in spring were tested for WNV and SLEV; none was positive (Reisen et al. 2010). Because the Coachella Valley represents a natural stopover site for migrants traversing the Salton Sea and adjacent desert, it could be one of the first locations in the western United States to have arboviruses introduced (or reintroduced) each spring from tropical regions to the south (Reisen et al. 2010). The lack of any viruses in arriving migratory birds, despite viruses such as SLEV and WNV being widespread in the area and often isolated from mosquitoes there, suggested that these arboviruses are not transported into southern California each spring by birds (Reisen et al. 2010). Similarly, >16,000 birds migrating into Hawaii were tested for active WNV infections, and none was positive (Kilpatrick et al. 2004). Because negative results may often not be published (Stallknecht 2007), the percentage of either migratory or resident birds with active viremia at any given time may be even lower than current data indicate (Table 1).

Infected birds found dead are hard to interpret because virus may have killed them before they moved any significant distance. Nevertheless, the data on dead birds positive for WNV RNA in tissues reported by the public in North America since 1999 also lead to a conclusion of mostly resident birds being associated with this arbovirus. For example, among the 12 bird species most frequently reported as WNV-positive in the United States from 2003 to 2004 (McLean 2006), only 3 were species that undertake regular migration, and only 1 of those—the American White Pelican (*Pelecanus erythrorhynchos*)—is a long-distance migrant. The three migratory species represented 3% of the 18,928 total dead birds reported (McLean 2006).

The same pattern of mostly resident species being associated with WNV was seen when the dead-bird data were analyzed separately for New York (Eidson et al. 2001), Colorado (Nemeth et al. 2007), and California (Reisen et al. 2009a, Wheeler et al. 2009). Of >1,100 dead birds of 76 migratory species tested in California, 11% were positive for WNV RNA, but only 0.7% of those that were found in spring during the migratory period were positive, compared with 8.4% of resident species during the same time (Reisen et al. 2010). The percentage of migratory species positive for WNV RNA in spring was significantly lower than that for migratory species during summer, prompting Reisen et al. (2010) to conclude that most of the WNV-positive migrant birds found dead were likely infected after they arrived in California.

INFECTED BIRDS TRAVEL FAR ENOUGH TO TRANSMIT VIRUS AT NEW SITES

If birds are to successfully transport arboviruses, they must encounter transmission-competent arthropod vectors that then take blood meals and become infected with virus, and this exposure to the vector must occur after bird movement begins or, in some cases (e.g., in intercontinental transport of virus), after the birds have traveled relatively long distances. Evaluating the likelihood of this happening requires information on (1) how long moving birds can be expected to be viremic, (2) how far birds travel while infected, and (3) whether there are transmission-competent vectors to which they can be exposed upon their arrival at new locales (Scott 1988, Owen et al. 2006).

Period between infection and cessation of viremia.—Most arboviruses cause viremia in their avian hosts relatively soon after initial infection and produce short viremia. For example, among alphaviruses, onset of viremia in captive studies of experimentally inoculated individuals is 1–5 days after infection, and viremia typically lasts 2–5 days, although titers are generally too low by days 4–5 to infect arthropods (Hammon et al. 1951, Bowen and McLean 1977, Dickerman et al. 1980, Hardy and Reeves 1990b, Lundström et al. 1993, Reisen et al. 2003c). In flaviviruses such as WNV and SLEV, experiments in a variety of species have shown onset of viremia 1–4 days after inoculation, with viremia lasting 1–6 days (Hammon et al. 1951; Work et al. 1955; McLean et al. 1983; Malkinson and Banet 2002; N. Komar et al. 2003; Reisen et al. 2003c, 2004b, 2005a; Langevin et al. 2005; Nemeth et al. 2006a, b). Thus, about the longest time between when a bird is bitten by an infected vector and when it can still transmit the virus to another vector would be 7–8 days, with a shorter period for most bird species and most arboviruses.

Travel distance during the period of infectiousness.—The short period, on average, that most birds are infectious to vectors puts constraints on how far any infected bird can potentially transport an arbovirus (McLean and Scott 1979; McLean and Bowen 1980; Scott 1988; Rappole and Hubálek 2003; Reisen et al. 2003a, c). Therefore, evaluating the likelihood of effective transport of virus over relatively long distances requires knowing how far birds typically move during the interval of ≤8 days between initial infection and cessation of viremia.

The prevailing assumption has been that infected birds can sometimes travel far enough in a short enough period to be effective arbovirus transport agents (Dickerman et al. 1980, McLean 2006, Owen et al. 2006). This belief is probably based, in part, on observations and inferences that landbirds routinely cross 1,000 km or more of inhospitable terrain such as oceans or the Sahara Desert nonstop in 60 h or less of flight time (Blondel 1972, Williams and Williams 1990, Delingat et al. 2008). For example, radar data have suggested that migrant birds leaving the eastern coast of North America may travel to Bermuda in 18 h, to the Caribbean in 64–70 h, and to the northern coast of South America in 80–90 h (Williams et al. 1978). If infected birds travel at these speeds, they would move fast enough and far enough for even intercontinental transport of arboviruses to occur. Radar data do have limitations, however: recent work, for instance, indicates that many trans-Saharan migrants do not cross the desert in nonstop flights as was previously assumed, which means that it may take most birds longer to make the crossing than had been thought (Schmaljohann et al. 2007a, b).

Generalizations are difficult because migratory species show enormous variation in how far they may travel during short time intervals. At one extreme, the Bar-tailed Godwit (*Limosa lapponica*) has been reported to fly the ~11,000 km between its Alaska breeding grounds and its winter range in Australia and New Zealand nonstop in 7–8 days (Gill et al. 2005). Other shorebirds also undertake long nonstop flights over water: for example, a Pacific Golden-Plover (*Pluvialis fulva*) flew the 3,900 km between Hawaii and the Alaska peninsula in <3 days (Johnson et al. 2004). A migrating Swainson's Thrush moved 1,512 km from east central Illinois to southwestern Manitoba in 7 days (Cochran 1987); thrushes average ~265 km day^{-1} (Wikelski et al. 2003, Stutchbury et al. 2009). Roughly similar distances moved per day have been reported for Barn Swallows (*Hirundo rustica*), Spotted Flycatchers (*Muscicapa striata*), and Willow Warblers (*Phylloscopus trochilus*) migrating between Africa and Europe (Alerstam and Lindstrom 1990). Purple Martins (*Progne subis*) average ~500 km day^{-1}; one was documented migrating 7,500 km between the Amazon basin of Brazil and northern Pennsylvania in 13 days (Stutchbury et al. 2009). These movements illustrate the capacity, in theory, for birds to transport arboviruses long distances.

However, other data indicate that passerine birds probably often move much shorter distances per day during migration. Estimated flight distances for 30 Eurasian passerine species varied from 27 to 75 km day^{-1} (Hildén and Saurola 1982, Alerstam and Lindstrom 1990), which means that these birds could maximally cover between 215 and 600 km during the interval of ≤8 days between infection and cessation of viremia. For 13 species migrating to tropical Africa in autumn, median duration was estimated as 88 days from departure to arrival, and for 30 species migrating within the north temperate zone, median durations were 32–42 days (Hildén and Saurola 1982, Alerstam and Lindstrom 1990). The available data on estimated daily migratory distances are of course based on birds for which virus infection status was unknown, and no studies have followed the movements of an arbovirus-infected migratory bird in the field (but for avian influenza virus, see van Gils et al. 2007, Kleijn et al. 2010). Birds undertaking long, nonstop (e.g., transoceanic) flights typically are in very good body condition, with heavy fuel (fat) loads (Bayly 2006, Delingat et al. 2008), and if virus infection

compromises a bird's ability to put on these fuel loads, it may not undertake such a flight.

Another consideration is that many migratory species frequently interrupt their migration for several days at stopover sites (Rappole and Warner 1976, Biebach et al. 1986, Moore and Kerlinger 1987, Chavez-Ramirez et al. 1994, van Gils et al. 2007, Stutchbury et al. 2009, Altizer et al. 2011), where they may feed to replenish fat stores or wait out bad weather. Presence at stopover sites may serve to reduce the distance traveled during the time interval between when a bird is infected and when it ceases to be infectious to vectors, while simultaneously increasing the chance that a migratory bird might acquire a local arbovirus infection at the stopover site.

Despite the potential constraints on an infected bird's making a long flight, the occasional introductions of South American arboviruses into the southern United States in spring (Calisher et al. 1971, 1974) likely were by birds that migrated directly across the Gulf of Mexico. These spring introductions may reflect the presumed tendency of birds in general to migrate in the spring more rapidly than in the fall (e.g., Chavez-Ramirez et al. 1994, Bauchinger and Klaassen 2005, Bächler et al. 2010), possibly because of the selective pressures associated with obtaining a summer breeding site as early as possible (Kokko 1999, Brown and Brown 2000). As warming global climate accelerates the spring migration speeds in some species (Hüppop and Winkel 2006), the possibility of spring transport of arboviruses by birds may increase. By contrast, reduced migratory activity in other species in response to warmer temperatures (Pulido and Berthold 2010) may serve to further decrease the likelihood of birds transporting arboviruses over long distances.

Exposure to vectors after moving.—Another major question is the extent to which transmission-competent arthropod vector species are active when birds reach their destinations and, thus, the extent to which these vectors might become infected by an arbovirus from a recently arrived bird. In temperate latitudes of North America and Europe, many migrant birds arrive at their breeding grounds or pass through in migration during the months of April and May, a time when the numbers of *Culex* and other transmission-competent mosquitoes are generally low at most locales (Madder et al. 1983, Scott 1988, Reisen and Reeves 1990, Sellers and Maarouf 1993, Crans et al. 1994). In the one case of JEV that was isolated from tissue of a migrant warbler in Japan, the isolation occurred 3 weeks before the earliest that the virus was found in mosquitoes (Takahashi et al. 1972). A similar situation held for a Gray Catbird found with EEEV in Louisiana prior to the appearance of the virus in local mosquitoes (Kissling et al. 1955). It seems unlikely that a viremic bird in these circumstances will initiate transmission, simply because the odds are low that it will be bitten by an ornithophilic mosquito (Sellers and Maarouf 1993). Low exposure to transmission-competent arthropod vectors (either because the competent vectors are not active then or because no competent vectors occur there) may help explain why some South American arboviruses, when brought to the United States by migrant birds in spring (Calisher et al. 1971, 1974), have not become established on this continent. For example, MAYV is transmitted in the tropics mostly by sylvatic mosquitoes not found in North America (de Thoisy et al. 2003, Weaver and Reisen 2010); this virus may have different intrinsic requirements for transmission or replication than are met by local Louisiana mosquitoes.

On the other hand, birds moving to warmer, more tropical areas (e.g., North American migrants in fall) will be more likely to encounter arthropods upon arrival at their destinations (or at stopover sites), because ornithophilic mosquitoes such as *Culex* can be active year-round in those locales (McLean and Bowen 1980, Yuill 1986, Tesh et al. 2004). The consequence could be that virus is more likely to be transported southward by migrant birds in fall as they approach warmer (more vector-rich) areas, and not as likely to be moved northward in spring as birds encounter colder environments where vectors are less numerous and climatic conditions less suitable for virus transmission or replication (Scott 1988). The finding of markedly more virus in southbound fall migrants than among spring migrants in the Mediterranean (Watson et al. 1972) and in eastern North America (Bond et al. 1972, Dusek et al. 2009) supports this idea. However, the only truly definitive evidence for birds transporting arboviruses between North and South America (Calisher et al. 1971, 1974) involves spring migrants only.

If a viremic bird moves to a site where transmission-competent mosquitoes are present and is clinically ill after arrival, it may be an especially good candidate to transmit virus to mosquitoes. This could occur if infectious, sluggish birds are less likely to engage in anti-mosquito defensive behavior (Edman et al. 1974, Scott et al. 1988) and thus more likely to be bitten or, if febrile, more attractive to mosquitoes. Consistent with this possibility, the presence of dead WNV-infected corvids at different sites in California was positively correlated with finding WNV-infected *Culex* mosquitoes at each of these sites during onset of a 2006 WNV epidemic (Nielsen and Reisen 2007). In a study designed to test experimentally whether mosquitoes are more attracted to virus-infected birds, no difference was found in attraction to infected versus uninfected adult House Sparrows (Scott et al. 1988, 1990). However, the viruses used in these studies (EEEV, WEEV, SLEV) did not cause clinical illness in adult House Sparrows and, thus, may not have influenced the birds' behavior. Similar experiments are needed with more pathogenic viruses (e.g., WNV; O'Brien et al. 2010b).

Scenario 2: Virus Recrudesces in Birds after They Move

Because some arboviruses are known to be maintained asymptomatically at chronic, low levels in their vertebrate hosts (Levine et al. 1994, Kuno 2001, Tesh et al. 2005), several workers have suggested that episodic virus recrudescence (recurrence of viremia) may account for some cases of virus transport to a site by birds. This might be especially likely as a mechanism for annual reintroduction of arboviruses to a site following a period of interrupted virus transmission during the winter months (Reeves 1974, McLean and Bowen 1980), which could occur even in the absence of much physical movement by the avian host (i.e., in nonmigratory birds).

For a previously contracted arbovirus that has been cleared by the immune system to reappear in the circulating blood of a bird, it must have persisted in the bird in an infective form (Levine et al. 1994, Reisen et al. 2003a). The recrudescing virus must then be present in blood at high enough titer to infect vectors, and it must evade the acquired immune system long enough to be imbibed by a transmission-competent mosquito vector.

The history of this idea can be traced to Reeves et al. (1958), who isolated arboviruses (WEEV) in tissue of seven birds (all resident species in California) from 55 to 306 days after experimental inoculation. Two of Reeves et al.'s (1958) isolations came from blood of a House Sparrow and a Brown-headed Cowbird (*Molothrus ater*), and these data suggested that birds might maintain latent infections. More recently, recrudescence has been used to explain the detection of virus outside of the period of typical vector activity (Crans et al. 1994); in one interesting case, EEEV was found in blood of a wintering Hermit Thrush (*Catharus guttatus*) in Louisiana on 19 March, 2 months before any virus was found in mosquitoes at the same site (Kissling et al. 1955).

Levine et al. (1994) suggested that arboviruses such as SINV could remain in the host by targeting neurons and establishing a persistent productive (nonlytic) form. However, virus reactivation apparently was detected only in immunodeficient mice (Levine et al. 1994), and the virus was not observed to escape the neurons to induce infection in healthy animals (Strauss and Strauss 1994). There is evidence of occasional relapse of persistent arbovirus infections in chronically infected cattle, small rodents, and snakes, sometimes with altered virus properties (Gebhardt and Hill 1960, Morris 1988, Reisen and Monath 1989, Kuno 2001). Infectious WNV can be shed in urine of hamsters for up to 8 months after infection, although the virus is cleared from the blood rapidly after initial infection (Tesh et al. 2005). However, no arboviruses are known to establish long-lasting infections that spread during intermittent bouts throughout a host's lifetime in the manner of herpesviruses, human and simian immunodeficiency viruses, hepatitis viruses, and human papillomaviruses.

In a multi-year field study of EEEV in both birds and mosquitoes in New Jersey, Crans et al. (1994) isolated virus from six birds in late May and June, 7–51 days before the first isolations in mosquitoes, with four of the five species involved being summer residents that had migrated into the area. The timing of the virus isolations suggested that they were not from recently arrived birds from farther south, and Crans et al. (1994) interpreted the isolations to reflect recrudescence of EEEV in birds that had maintained chronic, latent infections. A Gray Catbird that was seropositive in May was found to be viremic in June of the following year (Crans et al. 1994). A similar observation was recorded in Louisiana, in which EEEV was found in a catbird on 25 April, probably well after the bird arrived and about a month before the first virus isolations in mosquitoes (Kissling et al. 1955). The study by Crans et al. (1994) is frequently cited as an example of arbovirus recrudescence, although

the fact that hatching-year birds that were seropositive to EEEV were also found prior to the first appearance of the virus in mosquitoes suggests that other factors (e.g., different vectors, undetected mosquito activity) may have been responsible for these early-season infections. Despite Reeves et al.'s (1958) laboratory results, little field evidence of virus occurrence in winter was found for WEEV in California, where virus was isolated from only 2 White-crowned Sparrows out of 3,242 passerine birds of multiple species tested between October and February (Reeves 1990).

Because studying virus recrudescence in the field is problematic in most cases (ideally requiring the capture and recapture of birds known to have been viremic at a particular time), attempts have recently been made to demonstrate virus persistence experimentally. Reisen et al. (2001, 2003a, b, c, 2004a, 2006b) did a series of comprehensive inoculation experiments to test whether a range of species (primarily House Finches and House Sparrows, but >20 others) exhibited chronic infections of WEEV, SLEV, or WNV that could potentially recrudesce to the point of being infective at a later time. Birds were tested for evidence of viremia and seroconversion, and a subset of birds were chemically immunosuppressed (Reisen et al. 2003b, 2004a), a treatment that mimics the presumed stress of migration or reproduction that might lead to reemergence of viremia. Some previously infected birds were also challenged with virus to see whether initial antibody production provided permanent protection from infection (Reisen et al. 2003a). No birds were found with evidence of circulating virus in blood after challenge (other than a few short ephemeral viremias with SLEV; Reisen et al. 2001) or after immunosuppression, nor were mosquitoes feeding on these birds infected by virus. Although infectious SLEV and WNV were isolated from tissues (e.g., kidney, lung, spleen, brain) of experimentally infected birds more than 6 weeks after initial infection (after passage through a mosquito cell line), in only one case was virus found in blood (Reisen et al. 2003c). These data did not show any strong evidence for chronic infections that could recrudesce to the point of being transmissible (Reisen et al. 2001, 2003a, b, c, 2004a, 2006b). The same conclusions were reached in similar studies of WNV by Nemeth et al. (2009) on House Sparrows and Owen et al. (2010) on Gray Catbirds.

Several studies using RT-PCR have shown RNA of WEEV, SLEV, and WNV to persist in various avian tissues for up to 6 weeks or longer after experimental infection (Reisen et al. 2001, 2003b, c, 2006b; Nemeth et al. 2009), although only rarely has viral RNA been found in blood after the period of initial viremia ends. RNA of BCRV was found in House Sparrows 15 days after experimental inoculation and 12 days after the birds were no longer viremic by plaque assay (Huyvaert et al. 2008). EEEV from an American Goldfinch (*Carduelis tristis*) and a Ring-necked Pheasant (*Phasianus colchicus*) failed to show plaques on Vero cell assays (McLean et al. 1985), and we have detected BCRV RNA circulating in blood of wild-caught House Sparrows and Cliff Swallows that did not plaque on Vero cells (O'Brien et al. 2011).

There is not yet consensus on the biological significance of viral RNA in blood (or, in some cases, in arthropod vectors) detected by RT-PCR that does not exhibit cytopathic (plaque) growth in cell culture assays. Non-plaque-forming viral RNA has been interpreted variously as noninfectious virus particles, inactivated virus, noncytopathic RNA, replicative intermediate RNA, or nonintact RNA, or as due to methodological artifacts such as titers being too low to plaque or to the insensitivity of plaque assay (Barrera and Letchworth 1996; Letchworth et al. 1996; Reisen et al. 2001; Kramer et al. 2002; Lambert et al. 2003; Choi and Jiang 2005; White et al. 2005; Moore et al. 2007; Brown et al. 2010a, b; O'Brien et al. 2011). In general, whether such RNA represents (or can potentially represent, after recrudescence) functional virus is unknown, although noncytopathic viral RNA detected in arthropod vectors is known to become cytopathic relatively soon after the vectors take blood meals (Bailey et al. 1978, Korenberg 2000, Reisen et al. 2002a, Brown et al. 2010a).

Serological studies of arboviruses in birds sometimes use seroreconversion as evidence of recrudescence (Gruwell et al. 2000). In this scenario, a bird that was formerly infected and produced antibodies to the virus (seroconverted) is recaptured and found to be seronegative. When captured a third time, the bird is again seropositive (Gruwell et al. 2000, Garvin et al. 2004). These data could demonstrate recrudescence of a persistent viral infection after antibodies had waned. It is not surprising that data on seroreconversion are sparse, given the inherent difficulty in repeatedly recapturing the same individual birds throughout a season or in subsequent years, coupled with the low numbers of birds that test positive for evidence of virus exposure at any time.

In one field study of SLEV (Gruwell et al. 2000) in which >43,000 House Finches and House Sparrows were caught, 1% were positive for SLEV antibodies, but of the antibody-positive birds, only 2% showed evidence of seroreconversion. No field evidence of seroreconversion was found for either EEEV or HJV in 204 birds sampled multiple times by Howard et al. (2004).

Especially in light of the experimental studies that have not demonstrated frequent arbovirus recrudescence in birds to a level sufficient to infect arthropod vectors, it seems unlikely that recrudescence, paired with bird movement, can account for the bulk of the occurrences of arboviruses in new locales or their periodic recurrence at the same site. However, recrudescence of a latent infection has been shown experimentally for *Borrelia burgdorferi* spirochetes (that cause Lyme disease), in which migratory restlessness in Redwings (*Turdus iliacus*) led to a relapse of a chronic infection sufficient to cause detectable spirochaetaemia (Gylfe et al. 2000).

Scenario 3: Birds Move and Infect Others by Direct Transmission

If an arbovirus-infected bird is eaten by a predator or scavenger, the possibility exists for transmission of virus from infected tissues of the dead animal to the one that eats it. If the infected predator or scavenger then develops a viremia, it could transport and potentially transmit virus by moving to another destination. In these ways, an arbovirus could be spread from site to site among hosts, independent of direct arthropod involvement. Infectious (i.e., cytopathic) WNV, JEV, SINV, WEEV, and EEEV have been found in organ tissues of birds, some of which were collected during migration (Kokernot and McIntosh 1959, Pavri et al. 1972, Takahashi et al. 1972, Ernek et al. 1973, Morris et al. 1973, Milby and Reeves 1990). The major unresolved questions with this scenario are whether recently infected dead birds maintain a high enough concentration of virus in tissues (including blood) at the time of ingestion to orally infect a predator or scavenger, and whether a bird chronically infected at a low level (see above) maintains enough infectious virus in tissues besides blood for oral transmission to occur upon its being consumed.

Recent experiments have shown that WNV can be contracted by owls upon feeding on experimentally infected mouse carcasses, and viremia in the owl may sometimes result (Nemeth et al. 2006a, b). Whether WNV or other arboviruses can be transmitted from bird carcasses upon ingestion is unknown. Some surveys have found raptors (hawks and owls, including some that eat birds) relatively highly represented among WNV-positive specimens (Nemeth et al. 2006a, 2007; Wheeler et al. 2009), a finding consistent with virus transmission to predators via ingestion of (either bird or mammal) carcasses. Reporting rates for raptors are probably biased, however, because they are large, easily found by the public, and more likely than smaller species to be taken to bird rehabilitation facilities.

Evidence indicating lateral (contact) transmission of several arboviruses via oral or cloacal shedding (Holden 1955b; McLean et al. 2001; N. Komar et al. 2002, 2003; Banet-Noach et al. 2003; Nemeth et al. 2006a, 2009; Dawson et al. 2007; Huyvaert et al. 2008) also suggests that infected birds may potentially introduce virus to a new locale and infect others without arthropod involvement whenever they come into close contact with uninfected birds. This scenario has the same requirement discussed earlier that birds with relatively high virus concentration (most likely in tissue, saliva, or feces) undertake movement. Contact transmission is probably most likely in highly colonial or communally roosting birds, in which aggressive interactions (Still et al. 1987, Brown and Brown 1996) may result in bird-to-bird contact of saliva or when individuals encounter the fresh droppings of others. In communal roosts, for instance, individuals in the lower levels of roost trees are more likely to have their plumage soiled by birds above them (Yom-Tov 1979, Evans and Sordahl 2009). Although the available evidence indicates the potential for various arboviruses to be shed orally or cloacally and infect other birds, too few field studies have been done to evaluate how often this occurs in nature or the role of direct transmission in general in arbovirus ecology (Nemeth et al. 2009).

SCENARIO 4: TRANSPORT OF INFECTED VECTORS

Another possibility is that birds transport virus-infected arthropods from place to place (Yuill 1986, Björsdorff et al. 2001, Hubálek 2004). This is most likely for viruses transmitted by ectoparasitic arthropods such as ticks or bugs that often travel on their avian hosts (Hoogstraal et al. 1961). The tick-borne encephalitis viruses, WNV, some of the bunyaviruses such as BAHV, and various other tick-associated viruses have occasionally been isolated from birds (Table 2) or from ticks in bird nesting colonies (Johnson and Casals 1972, Pavri et al. 1972, Watson et al. 1972, Yunker et al. 1972, Berge 1975, Yunker 1975, Hayes 1989, Gould et al. 2004, Hubálek 2004). Only a few isolations of viruses have been made from ticks collected directly off birds (Ernek at al. 1968, Converse et al. 1974), and most of those birds were not definitively known to be migrating at the time. By contrast, *Borrelia* spirochetes (and other bacteria) have been isolated relatively frequently from ticks actively traveling on migrating birds (Smith et al. 1996, Rand et al. 1998, Reed et al. 2003, Hubálek 2004, Morshed et al. 2005, Wright et al. 2006). As in the case of viremic birds, the finding of birds carrying infected arthropods is difficult to interpret because usually we cannot know the status of the bird (migrating or not), although the hitchhiking of ectoparasites may be less likely to affect a bird's migratory behavior than its being infected by a virus.

The best-studied example of arboviruses being transported via infected arthropods (or being transported at all) is that of Buggy Creek virus (Brown et al. 2007, 2008). The cimicid Swallow Bug that transmits BCRV resides year-round in the mud nests of Cliff Swallows (Moore et al. 2007; Brown et al. 2009a, 2010b). Cliff Swallows nest in colonies on bridges and highway culverts, and up to 6,000 nests sometimes occur on a single bridge. Swallows often carry bugs on their feet when the birds move between nesting colonies (Brown and Brown 2004, 2005). Transport of infected bugs seems to be the principal way that BCRV moves from colony to colony (Brown et al. 2008), given that infectious BCRV has not been isolated from any adult bird to date (O'Brien et al. 2011). Brown et al. (2007, 2008) used a long-running mark–recapture study that involved >100,000 captures of banded Cliff Swallows to estimate the likelihood of an individual moving from one colony to another per 2-day interval across the 2-month nesting season for a cluster of colonies in southwestern Nebraska along the North and South Platte rivers. The prevalence of BCRV in bugs at a given colony site was strongly correlated with both the likelihood of a Cliff Swallow immigrating into that site from other colonies in the study area and with the total number of transient Cliff Swallows moving through the site (Fig. 2A; Brown et al. 2007).

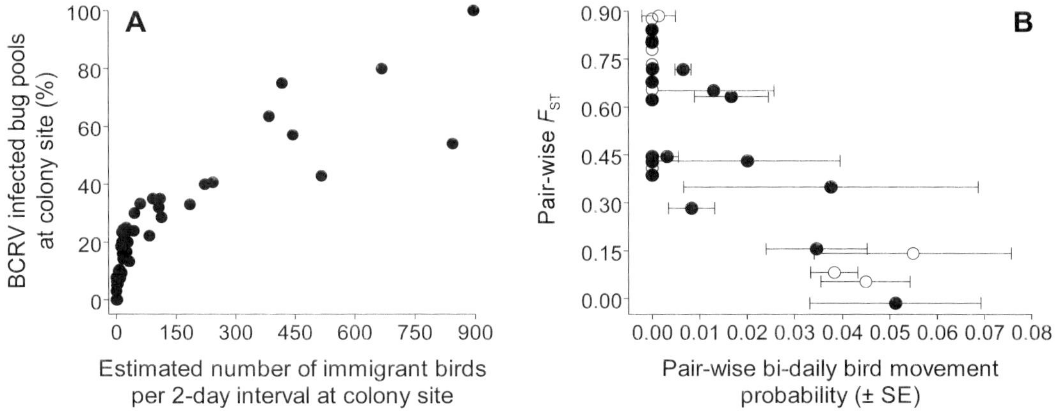

FIG. 2. (A) Percentage of Swallow Bug pools positive for Buggy Creek virus at a Cliff Swallow colony site in relation to the estimated number of immigrant Cliff Swallows moving into the site per 2-day interval throughout the summer nesting season. The number of immigrant birds was estimated from mark–recapture data. Infection of bugs increased significantly with the number of immigrant birds passing through a site (from Brown et al. 2007). (B) Genetic similarity (as measured by pairwise F_{ST} values for the E2 gene) of Buggy Creek virus per pair of Cliff Swallow colony sites in relation to the likelihood of a Cliff Swallow moving between each pair of sites per 2-day period during the summer season, ψ ($\pm$ SE), as determined from mark–recapture data. Two different virus lineages are shown by closed ($\bullet$) and open ($\circ$) circles, respectively. Genetic divergence between any two sites declined as bird traffic between them increased (from Brown et al. 2008).

Genetic Evidence for Arbovirus Transport by Birds?

The development of molecular sequencing techniques has allowed phylogeneticists to identify virus genetic structure and make inferences about geographic and temporal origins of individual virus strains. The prevailing view that wild birds are responsible for frequent transport of arboviruses has been supported by phylogenetic evidence of genetic similarity among viruses in different geographic regions. Genetic homogeneity (i.e., lack of spatial structure) of viruses is thought to result from frequent mixing of virus haplotypes, with migratory birds assumed to be the most likely agents responsible for introduction and reintroduction of different haplotypes into given areas (Scott et al. 1994, Weaver et al. 1994, Cilnis et al. 1996, Norder et al. 1996, Sammels et al. 1999, Powers et al. 2001, Nga et al. 2004, Weaver 2006, Brown et al. 2008, Young et al. 2008, Padhi et al. 2011).

Alphaviruses

Examination of genetic relationships among North American EEEV isolates (Weaver et al. 1994) showed a strong relationship between isolates from Michigan and Mississippi and another grouping of isolates from mostly the northeastern United States. The first group of genotypes was suggested to have evolved regionally, with virus being carried north–south by birds migrating along the Mississippi and Ohio River valleys (Weaver et al. 1994), and the second group is consistent with migratory bird corridors along the East Coast. North American EEEV isolates tend to group by year of isolation, which suggests a high degree of genetic conservation and led to the suggestion that the limited spatial structure is generated by these viruses being transported by birds (Brault et al. 1999, Powers et al. 2001, Weaver 2006, Young et al. 2008). Consistent with this interpretation, EEEV is apparently more associated with potentially migrating birds than any of the other North American arboviruses (Table 2).

When isolates of HJV collected between 1952 and 1994 were sequenced, they were found to be highly conserved, with little nucleotide divergence temporally or geographically (Cilnis et al. 1996). The slow evolutionary rate, combined with evidence that HJV consists of a single predominant lineage, indicated little genetic heterogeneity or spatial structure in this arbovirus. The lack of genetic diversity was assumed to arise because birds transport HJV between geographically distinct transmission foci, leading to mixing of genotypes and competitive exclusion where genotype mixing occurs (Cilnis et al. 1996). This hypothesis is consistent with the findings of a few migrant birds with HJV during fall (Kissling et al. 1955, Stamm and Newman 1963, Lord and Calisher 1970), although it appears that most birds in which HJV has been isolated in the field were not migrants at the time and, thus, not likely to be transporting virus (Table 2).

Phylogenetic examination of WEEV lineages using isolates from both South and North America revealed a lack of genetic diversity among most of them, regardless of the geographic location of isolation (Weaver et al. 1997). The genetic homogeneity was thought to reflect viral mixing and supported the hypothesis that birds transport WEEV both locally and intercontinentally on a frequent basis (Scott et al. 1994, Weaver et al. 1997). On the other hand, a more localized molecular analysis of WEEV isolates from California suggested that virus there has remained relatively isolated in distinct foci, with little regional (and no intercontinental) exchange between them (Kramer and Fallah 1999). Substantial transport of WEEV by birds at any scale seems unlikely, however, given how few migrants have been found with viremia; more than 92% of all known isolates of WEEV from birds in the field have come from nestlings (Table 2).

Phylogenetic analyses reveal that strains of SINV from South Africa are most closely related to isolates from northern Europe, with intercontinental transport by birds suggested as the most likely mechanism causing this result (Shirako et al. 1991, Norder et al. 1996). Furthermore, within Australia, genotypes cluster temporally, with isolates from widely separated locations showing identical sequences. Movement by birds was thought to have caused this pattern by disseminating SINV widely over the continent (Sammels et al. 1999). There are few records anywhere, however, of SINV being isolated from birds that were migrating (Table 2).

The strong correlation between the extent of genetic similarity in BCRV isolates from different bird nesting colonies and the extent of bird movement between the sites (Fig. 2B; Brown et al. 2008, Padhi et al. 2011) supports the view that genetic homogeneity of virus can result from bird-mediated dispersal, at least on a local level. When BCRV isolates from different Cliff Swallow colony sites were sequenced and measures

of F_{ST} (an indication of genetic similarity among isolates) computed for pairs of colony sites and compared with the likelihood of Cliff Swallows moving between those specific sites, as estimated from mark–recapture data, there was a strong relationship between pairwise virus genetic similarity and extent of Cliff Swallow movement (Fig. 2B; Brown et al. 2008). Sites with more bird movement between them had virus that was more similar genetically than that among pairs of sites with limited or no bird movement. These results are probably the strongest indication we have for any arbovirus that the magnitude and direction of daily bird movement in a local area can accurately predict transmission foci and directly affect spatial genetic structure of arbovirus isolates. Note that in this case birds transported virus solely by carrying the infected arthropod vectors (Brown et al. 2008).

Flaviviruses

Analysis of the genetic structure of WNV in the Old World revealed genetic homogeneity between WNV strains in Africa and western Europe (Berthet et al. 1997, Burt et al. 2002, Charrel et al. 2003, Zeller and Schuffenecker 2004). The assumption has been that this similarity is caused by frequent virus interchanges between the continents, probably brought about by migratory birds (Hayes 1989, Berthet et al. 1997, Lundström 1999, Charrel et al. 2003, Hubálek 2004, Kramer et al. 2008). Because Israel is a natural migratory funnel between the Mediterranean and the African Rift Valley, with many birds stopping over there (Malkinson and Banet 2002), intercontinental virus transport might be most likely to be detected there. The finding of WNV-infected migratory storks in Israel (Malkinson et al. 2002) is consistent with at least occasional transport of this arbovirus into Africa from the north through the eastern Mediterranean by birds (although that storks transported virus in that case was not established definitively; see above).

Phylogenetic studies of WNV in North America showed strong genetic homogeneity among isolates taken from different locales relatively soon after the initial outbreak (Bertolotti et al. 2007, Brault et al. 2007); judging from the genetic studies of other virus species, this would seem to implicate birds as transport agents (Weaver 2006). However, since its initial introduction and spread across North America, WNV has shown an increasing genetic spatial structure, indicating establishment of the virus in local enzootic cycles (Davis et al. 2003, 2005; Estrada-Franco et al. 2003; Bertolotti et al. 2008; Grinev et al. 2008), presumably without frequent introductions from outside a local transmission zone. If birds were involved in spreading the virus initially, their involvement recently would seem to have waned as WNV becomes more genetically differentiated at the local level.

The establishment of local genetic structure in WNV may mirror that of the other predominant flavivirus in North America, SLEV. Genetic studies indicate that SLEV persists with little change between consecutive years, but then becomes locally extinct before new genotypes are introduced or arise through local evolution (Kramer et al. 1997, Chandler et al. 2001, Reisen et al. 2002b). Birds are one potential mechanism to account for the introduction of virus into new areas, but SLEV persistence through vertical transmission among mosquitoes and overwintering of infected adult mosquitoes is possible as well; none of the proposed mechanisms has solid empirical support (Kramer et al. 1997, Reisen et al. 2002b). The strong association between SLEV and resident birds and nestlings (Table 2) is consistent with its localized population genetic structure.

JEV exists as several distinct genotypes (Nga et al. 2004), two of which are widely distributed from India across Southeast Asia (e.g., Vietnam) to East Asia (e.g., Japan, Korea) (Burke and Leake 1988, Erlanger et al. 2009, Weaver and Reisen 2010). The virus's genetic similarity over this wide area suggests frequent introductions between geographic regions, and some evidence now suggests that one of the genotypes has recently become more widespread (Nga et al. 2004). Transport of JEV by birds is one hypothesized mechanism to account for this genetic pattern (Nga et al. 2004, van den Hurk et al. 2009), although the virus has relatively rarely been found in wild birds, the primary mosquito vector prefers mammals (Weaver and Reisen 2010), and the known field isolates from birds all came from nestlings (Table 2).

Birds are thought to be responsible for relatively recent introductions of Usutu virus (USUV), a flavivirus related to JEV, to Europe (Weissenböck et al. 2002). This virus is found in tropical and subtropical Africa and, until the introduction to Austria, had not been recorded outside of Africa. In late summer 2001, a sudden die-off of Eurasian Blackbirds (*Turdus merula*), Great

Gray Owls (*Strix nebulosa*), and Barn Swallows in and around Vienna was attributed to USUV. Phylogenetic analysis showed that isolates taken from dead birds in Austria were closely related to strains from South Africa and, in fact, appeared to be identical at the amino acid level (Weissenböck et al. 2002). Barn Swallows were suggested to be responsible for introduction of USUV, owing to their wintering in South Africa and the detection of USUV RNA in a dead Barn Swallow. However, USUV apparently causes heavy mortality in Barn Swallows (Weissenböck et al. 2002), and thus it is unclear whether this species (or any bird) could have naturally introduced it from such a long distance.

COMPARISON TO A MAMMAL-ASSOCIATED ARBOVIRUS

If arboviruses amplified and transported by birds show genetic homogeneity over large geographic distances because of repeated reintroductions that result from widespread virus dispersal, we would predict that ecologically similar viruses not associated with mobile hosts such as birds should show more heterogeneity in genetic structure. La Crosse virus (LACV) is a mosquito-transmitted bunyavirus that is amplified primarily in sedentary arboreal rodents (chipmunks and squirrels); birds have no known role in its transmission. It occurs in temperate latitudes of eastern North America, largely overlapping portions of the range of both EEEV and HJV. Although several distinct lineages of LACV circulate in North America, sometimes sympatrically, LACV shows genetic homogeneity over wide areas, with virus from Missouri north to Minnesota being genetically similar (Armstrong and Andreadis 2006). In many ways, its spatial structure closely resembles that of EEEV. The LACV example illustrates that reduced genetic structure of arboviruses in temperate regions can occur in the absence of bird involvement and underscores the need to evaluate alternative hypotheses (see below) to explain the population genetics of EEEV, HJV, and other temperate-latitude arboviruses.

LOCAL TRANSPORT

Given that arboviruses in general have been isolated more frequently from resident birds than from migratory species (Table 2), local virus transport by birds in and around their nesting territories may be ecologically and epidemiologically more relevant than that occurring over longer distances. About 85% of bird species in the world are believed to be territorial to some extent, at least during the breeding season (Lack 1968), and, thus, to confine their activities to relatively small areas; consequently, most birds undergo area-restricted movement during much of the summer. The times when territorial birds are defending territories and raising nestlings, and the subsequent addition of dispersing fledglings to the local population in late summer (Unnasch et al. 2006b), may often overlap with periods of high vector activity and, thus, the principal transmission season of some arboviruses. The consequence is that birds may most commonly transport viruses over short distances roughly equivalent to the longest axis of a typical territory.

The east-to-west, concentric spread of WNV in North America is more consistent with local movement of resident species such as House Sparrows than with movement of migratory species (Rappole and Hubálek 2003, Rappole et al. 2006). Even for arboviruses such as VEEV in which birds are relatively insignificant amplifiers, they still may be responsible for local spread of virus whenever individuals (during the short time they are viremic) undertake short-distance, local movements (Bowen and McLean 1977). In general, local bird movement may result in rolling "waves" of arbovirus infection that ripple over short geographic distances, although exactly how this happens and by which birds is unknown. One scenario is suggested by the study of radiotagged American Crows naturally infected with WNV (Yaremych et al. 2004, Ward et al. 2006) and a spatial analysis of mosquito infection in relation to where dead crows were found (Reisen et al. 2006a): after a crow is initially infected at a roost site, it moves on the first day to feed (before it becomes too sick to move), becomes ill at the edge of its foraging range or territory, and infects a vector there before it dies (Nasci et al. 2002, Nielsen and Reisen 2007). The arthropod (mosquito) then bites another bird in an adjacent territory or roost. This chain of events requires adjacent territories or foraging ranges that span a mosquito's movement radius. Virus infection of raptors or scavengers via bird carcass consumption might also move WNV in localized waves.

EEEV is known to occur in spatially discrete foci centered on swamps where the principal mosquito vector, *Culiseta melanura*, is abundant (Morris et al. 1980, McLean et al. 1985, Morris 1988, Crans et al. 1994, Howard et al. 1996). Amplification of this virus occurs by resident birds within or near each swamp (Dalrymple et al. 1972, Emord and Morris 1984, Unnasch et al. 2006b). Resident birds undoubtedly transport virus over short distances within swamps, but despite the frequency with which EEEV has been isolated from birds (Table 2), transient migratory birds are thought not to have much role in either exporting or annually reintroducing it to its swamp-associated spatial foci (Emord and Morris 1984, McLean et al. 1985). A similar argument was made for SLEV (Day and Stark 1999).

Even if birds are responsible for occasionally exporting arboviruses such as EEEV or SLEV from local foci, not all birds of a given species may be equally likely to do so. The relatively local spread of BCRV among Cliff Swallow nesting colonies is perpetrated, in part, by birds whose nesting attempts fail, which thus disperse to new colonies to renest, or by individuals that are nonbreeders (Brown et al. 2008). This suggests that local transport of virus may sometimes be predicted by avian nesting success, with successful birds less likely to move (and spread virus) within a summer. The consequence is that arboviruses may be dispersed more widely by birds whenever environmental conditions (including local virus amplification and subsequent mortality in nestlings) lead to widespread nesting failures in a given area.

Transport of viruses on a mostly local scale also follows from the fact that more sedentary nestling birds are often better amplification hosts for arboviruses than adults (O'Brien et al. 2010a, b, 2011). A striking pattern among the reported cases of arbovirus isolations from wild birds worldwide is that ~64% have come from nestlings (Table 2; many of these are of WEEV and BCRV). That over half of all known arbovirus detections in birds in the field have come from nestlings, despite the relatively limited sampling of them (Table 1), suggests that they may be critical in the transmission dynamics of some arboviruses (Burkett-Cadena et al. 2010, O'Brien et al. 2010b). Mosquitoes (Blackmore and Dow 1958, Scott et al. 1990) and Swallow Bug vectors (C. Brown pers. obs.) at times prefer to feed on nestlings, probably because they are too young to exhibit anti-arthropod defensive behavior (Scott et al. 1988; cf. Griffing et al. 2007). Studies showing that nestlings often have high viremia levels that exceed those of adults (Holden et al. 1973, Bowen et al. 1980, Ludwig et al. 1986, Scott et al. 1988, Mahmood et al. 2004a, O'Brien et al. 2011), and the high seroprevalence sometimes seen in hatching-year birds (Howard et al. 2004, Hamer et al. 2008), also suggest that nestlings frequently amplify arboviruses. Furthermore, even if mosquitoes do not exhibit an age-related preference and instead feed on adults and nestlings only in proportion to their relative abundance in the population, the greater numbers of nestlings than of adults during a summer suggest that more nestlings than adults will be infected (Burkett-Cadena et al. 2010). Thus, the extent to which nestlings fledge and move while still viremic will determine in part the extent to which arboviruses may be transported on a local scale.

No data are available to evaluate whether birds ever fledge while viremic, although in most species recently fledged young (that may still be dependent on their parents) probably do not travel very far. For example, Purple Martins rarely move farther than ~1 km from their natal nest during the first 4–5 days after fledging (Brown 1978), Gray Catbirds stay within 18 m of their nest for 5–8 days after fledging (Zimmerman, in Cimprich and Moore 1995), White-crowned Sparrows are sedentary in or near their natal territory for 1–2 weeks after fledging (Chilton et al. 1995), and Mourning Doves (*Zenaida macroura*) remain within 45 m of their nest tree for 12–15 days after fledging (Mirarchi and Baskett 1994). Many other species show similar patterns.

Even if nestlings remain relatively sedentary, their ability to amplify virus may still contribute to virus spread, if mosquitoes that feed on them subsequently feed on more mobile adults or juveniles associated with a nesting territory or on migrants passing through. Although many North American birds, especially Neotropical migrants, may nest largely before the peak (late-summer) season for arbovirus transmission by mosquitoes, up to 25% of terrestrial bird species are known to breed at least occasionally in September or later (Koenig and Stahl 2007). Late nests may be important in arbovirus transmission cycles (Buescher et al. 1959, O'Brien et al. 2010b), especially because they occur at the same time that some birds have started migration and when transmission-competent mosquitoes such as *Culex* are still active. In general, however, we know relatively little about the true extent of late nesting in most bird populations (Koenig and Stahl 2007). Claims

that nestlings are not important hosts for some arboviruses (Loss et al. 2009) are premature in the absence of thorough surveys of late nesting by competent host species. Modeling by Unnasch et al. (2006b) showed that the appearance of hatching-year birds in late summer has a greater effect on promoting enzootic outbreaks of EEEV in swamps than the presence of uninfected adult birds.

The concentration of birds like Cliff Swallows into large breeding colonies, or of species such as Purple Martins, American Crows, American Robins, European Starlings, and various icterids into roosts in late summer, may facilitate mosquito feeding because of the density of potential hosts; for example, mosquitoes seem attracted to larger Cliff Swallow colonies (Brown and Sethi 2002). The consequence is that highly social birds may be more likely to transport arboviruses over short distances between their nesting colonies or roosts and their foraging sites simply because they are more likely to be infected than more solitarily living species (Brown et al. 2001). Colonial or communally roosting species are also less likely to be territorial and thus prone to wander more freely. For example, some evidence indicates that colonial waterbirds such as herons and egrets are effective amplifying hosts for some arboviruses (Kissling et al. 1954, Buescher et al. 1959, Boyle et al. 1983, McLean et al. 1985, Gottdenker et al. 2003, Unnasch et al. 2006a; cf. Reisen et al. 2009b), and these species are well known to undertake both short- and long-distance movements in late summer (Scherer et al. 1959, Oberholser 1974, Bowen and McLean 1977) at a time when arboviruses are most likely to be transmitted between mosquitoes and vertebrate hosts.

Conclusions

Relatively few wild birds maintain viremia sufficient to infect arthropod vectors with virus at any given time (Table 1). The chain of events leading to these birds' successful transport of arboviruses in space represents a series of probabilities that must be high enough to have a measurable likelihood of happening (Owen et al. 2006); in short, these events must converge to present "the perfect storm" for each occurrence of successful virus transport and establishment at a new locale. Presumably, the probability of these events happening in the required sequence (initial infection, travel, viremic enough on arrival to infect a transmission-competent vector) is, in general, inversely related to the distance a bird travels and how long it takes to get there. The probability of successful arbovirus transport by birds also depends heavily on the likelihood of an infected bird encountering a transmission-competent vector after it moves.

Kilpatrick et al. (2004) modeled the likelihood of viremic migratory birds introducing WNV to Hawaii, and even though forced to make a number of perhaps generous assumptions about the likelihood of the components in the specific chain of events occurring, they concluded that the chance of wild birds introducing this arbovirus was almost nil. Nevertheless, the impression is engrained among arbovirologists and public health workers that bird transport of virus occurs regularly because of the vast numbers of birds and mosquitoes out there, and by the idea that "it takes just one" to introduce or reintroduce virus to a specific locale. Two recent studies are the latest examples of the conventional wisdom. Balança et al. (2009) found no evidence of past WNV exposure in >1,100 migratory birds sampled in southern France yet qualified their conclusion of no role for birds in virus transport by referring to the "millions of migratory birds visiting this region," implying that some of these might nevertheless introduce it. (This logic seemingly suggests a predetermined outcome, regardless of the empirical data obtained.) Dusek et al. (2009) detected WNV in 17 birds (0.8% of those) sampled for WNV in fall 2003 in the eastern United States (as noted above, most could not be distinguished as either migrants or local residents) and concluded that this was "a remarkable percentage considering that millions of migratory birds traverse the Atlantic and Mississippi flyways in the fall"; the authors implied that this indicates large-scale virus transport by birds.

This sort of thinking is easy to understand. For example, for the Gray Catbird, the most commonly tested species for WNV in Dusek et al.'s (2009) North American survey, virus was isolated from 8 of 3,178 birds (0.252%). If we use Partners in Flight's (2007) population size estimate of 10 million catbirds in North America (despite wide uncertainty in this estimate; Thogmartin et al. 2006), we might infer that >25,000 Gray Catbirds are viremic for WNV in fall at any particular time. This certainly indicates the capacity of this species to commonly amplify WNV in North America, and because of the timing of the isolations, some might conclude that many of the viremic birds were migrating. However, as we emphasized earlier, we do not really know whether those eight birds were in fact migrants at the time of sampling, or if they were migrants, whether they were infected locally and the extent to which an infected migrant is likely to continue to move. If none of the eight birds were actual migrants or did not continue to migrate once infected, then an equally plausible inference is that no viremic Gray Catbirds were migrating. The available data only allow us to conclude that between zero and 25,000 migrant Gray Catbirds might be transporting virus at any particular time in the fall in North America, and we have no basis to judge which estimate is closer to reality.

ASSUMING THAT BIRDS TRANSPORT ARBOVIRUSES MODERATE TO LONG DISTANCES IS PREMATURE

Although birds likely transport arboviruses over substantial distances in some instances (Calisher et al. 1971, 1974), the empirical evidence for wild birds playing a significant role in the continental or intercontinental transport of arboviruses is at best anecdotal and not sufficient, in our judgment, to conclude that bird-mediated transport is important in virus maintenance, transmission, and evolution at any but local spatial scales. We suggest that the notion that birds "must" be responsible for virus transport because there is no strong evidence implicating other dispersal mechanisms is contrary to established procedures of scientific verification and inference. In light of current evidence, we see the following as the strongest arguments against assuming that birds are the primary long-distance transport agents of arboviruses: (1) the lack of behavioral information on how viremia influences the movement activities of birds in nature (and the limited evidence suggesting that infection can sometimes reduce movement propensity); (2) the low frequency with which arboviruses have been isolated in birds definitely known to be migrating; (3) our not knowing where or when the migrant birds that have been found with virus were infected; (4) accumulating evidence from serological surveys suggesting that migratory species either are less often infected with arboviruses or, if infected, perhaps more likely to succumb than resident birds; and (5) the experimental data showing clearly that arbovirus recrudescence is unlikely.

The strongest suggestive evidence for birds being important in the transport of arboviruses comes from the molecular data showing genetic similarity of bird-associated viruses in different geographic locations that are connected by avian migratory routes (Weaver et al. 1994, Cilnis et al. 1996, Young et al. 2008) and the empirical relationship between bird movement and genetic similarity of virus at the local level (Brown et al. 2007, 2008). Nevertheless, it seems that the role of birds has been somewhat uncritically invoked to explain even the genetic data without full evaluation of alternatives. The genetic similarity of the mammal-amplified LACV over wide geographic areas demonstrates that bird involvement is not necessary to produce homogeneity of virus spatial structure. Could slower rates of virus evolution in temperate latitudes, owing to colder temperatures and shorter transmission seasons (Cilnis et al. 1996, Weaver et al. 1997), and the restriction of many arboviruses to a single mosquito vector species, account in part for lower arbovirus spatial structure in North America, irrespective of what host amplifies the viruses?

We are not the first to question whether wild birds regularly transport arboviruses over long distances. Takahashi et al. (1972) and Reisen et al. (2003c, 2010) doubted the importance of migratory birds in transporting arboviruses because of the infrequency with which virus (or antibody) was detected in migrants. Morris (1988) argued that migratory birds probably do not serve as sources for new outbreaks of EEEV, because that scenario would predict a spread of EEEV activity from north to south in fall as birds migrate southward. Such a diffusive spread is typically not seen; instead, EEEV occurs in the same foci repeatedly where the appropriate mosquito vectors occur. Scott (1988) questioned whether arboviruses are likely to be transported north in the spring, presumably because of the lack of widespread transmission-competent mosquito activity when spring migrants are moving north, and suggested that virus transport southward in the fall might be more likely. The hypothesis (Scott 1988) that virus might be maintained in semitropical areas such as Florida and then transported northward slowly by migrating birds or by local bird movements has been embraced by some, but the reports of arboviruses in birds in Florida come mostly from territorial, resident species or fall migrants, so significant northward movement within North America seems unlikely.

ALTERNATIVE HYPOTHESES

Clearly arboviruses do move in space. The most publicized recent example is the introduction to North America of WNV, most likely (according to molecular evidence) from the Middle East or Africa (Charrel et al. 2003). Although there are several ornithological scenarios in which a viremic wild bird from the Middle East or Africa might make it to New York City, none is very likely (Rappole et al. 2000). Introduction via a mosquito in an airplane or ship seems more plausible, given that infected vectors such as mosquitoes are known to disperse arboviruses over wide areas by anthropogenic means (Reisen et al. 1972, 2004b; Yuill 1986; Zeller and Schuffenecker 2004; Kilpatrick et al. 2006a; Tatem et al. 2006; Weaver and Reisen 2010). Many arthropod vectors can maintain permanent virus infections as long as they live and, thus, can potentially undergo longer periods in transit to new locations than birds whose viremias and infectious periods are short-lived.

Even in the absence of anthropogenic transport, vectors may move long distances. For example, a genetic analysis of mosquitoes (*Culex tarsalis*) that transmit WNV in the western United States suggested that widely separated mosquito populations are more genetically homogeneous than is usually assumed (Venkatesan and Rasgon 2010). If such genetic homogeneity reflects mixing of haplotypes (as assumed by some for arboviruses; see above), this would suggest that these mosquitoes regularly disperse long distances and, thus, could move WNV in the process (Venkatesan and Rasgon 2010).

Mosquitoes in Australia are blown by winds over distances of up to 648 km night^{-1}, and dispersal of infected mosquitoes may account for the movement of JEV from Papua, New Guinea, to mainland Australia (Hanna et al. 1996, Kay and Farrow 2000, Ritchie and Rochester 2001, Johansen et al. 2004). Analysis of wind trajectories shows that WEEV-infected *Culex* can be carried progressively north from southern Texas as far as Minnesota and North Dakota and possibly as far as Manitoba (Sellers and Maarouf 1988, 1993). EEEV could be transported by infected mosquitoes from North Carolina north and east as far as upstate New York and Michigan (Sellers and Maarouf 1990). Wind-dispersed *Culex* may have introduced Rift Valley fever virus to Egypt from Sudan (Sellers et al. 1982). These results suggest that some insect vectors have greater potential to move over long distances than is generally assumed (Johnson 1969, Pedgley 1983, Drake and Farrow 1988, Sellers 1989, Kay and Farrow 2000, Goldberg et al. 2010), and thus more research should be directed at measuring long-range dispersal of arboviruses by arthropods. On the other hand, viruses could be less likely to be transported by arthropods if infection reduces the vector's survival, as has been demonstrated in some mosquitoes for EEEV and WEEV (Scott and Lorenz 1998, Lee et al. 2000, Moncayo et al. 2000, Mahmood et al. 2004b).

The popularity of the idea that virus transport (or virus recrudescence) by birds explains, in particular, the annual recurrence of arboviruses at specific foci may derive from the general lack of understanding of how most arboviruses overwinter in temperate latitudes. Because vectors are not numerous and virus is rarely found in adult vectors in winter (Rush et al. 1963, Rosen 1987, Reeves 1990, Reisen et al. 2006b), most workers have assumed (in the absence of reintroduction by birds) that arboviruses survive the winter period of interrupted transmission by overwintering in diapausing adult vectors or by being transovarially (or otherwise vertically) transmitted from infected females to eggs that overwinter locally (Reeves 1974, Scott and Weaver 1989, Reisen 1990, Crans et al. 1994). A number of arboviruses have been shown to be vertically transmitted in vectors (Watts et al. 1973, Watts and Eldridge 1975, Mims 1981, Rosen 1981, Turell 1988, Goddard et al. 2003, Anderson and Main 2006), and there are some reports of arboviruses surviving in adult arthropods for more than one year (Brown et al. 2010a). Although some of the bird-associated alphaviruses (e.g., EEEV, WEEV, and HJV) do not seem to be vertically transmitted routinely (Henderson and Brust 1977, Morris and Srihongse 1978, Sprance 1981, Tesh 1984, Reisen et al. 1996; cf. Brown et al. 2009b), nevertheless more attention should probably be given to the extent to which temperate-latitude arboviruses overwinter locally in surviving adult vectors of known or unknown species or in their eggs.

We have not attempted here an exhaustive, or necessarily even a critical, review of the alternative ways that arboviruses may be dispersed in space, and these alternative hypotheses also lack strong empirical support in most cases. Our intent is simply to point out that these other mechanisms are plausible, understudied, and deserving of greater attention by field workers in the future.

Future Research

Because there is empirical evidence that some birds can transport arboviruses locally (Ward et al. 2006; Brown et al. 2007, 2008), workers should probably emphasize study of how local bird movement contributes to the persistence and possible spread of arboviruses within and between given foci. The idea (Bowen and McLean 1977, Rappole et al. 2006) that virus can be transported across a landscape through concentric, outward radiation of infected birds moving locally over relatively short distances (perhaps in late summer as territorial defense ceases and birds begin postmigratory wandering) deserves more attention. Field data on typical daily movement distances of model species, such as House Sparrows or American Robins, could be combined with modeling (e.g., Okubo 1980, Shaman 2007) to evaluate the potential spread of arboviruses such as WNV in space and time.

We need additional research on how infection with different arboviruses specifically affects movement behavior of birds in both the field and laboratory; those of Lindström et al. (2003), Yaremych et al. (2004), Owen et al. (2006), van Gils et al. (2007), and Kleijn et al. (2010) provide useful starting points and models. Satellite-based technology for following larger species of migrant birds is now available and widely used (Pierre and Higuchi 2004, van Gils et al. 2007), and these methods plus those now being developed for tracking of even smaller (e.g., thrush-size) birds (Cochran and Wikelski 2005, Stutchbury et al. 2009, Bächler et al. 2010) could be employed to get a better understanding of how far birds typically travel per day while undertaking either local or long-distance movements. Although there are veterinary and public-health concerns associated with experimentally infecting birds with virus and releasing them (Enserink 2009), those found to be naturally infected (e.g., Table 1) could be monitored with tracking devices to determine directly how virus infection affects movement behavior in the field (Yaremych et al. 2004, Ward et al. 2006, van Gils et al. 2007, Kleijn et al. 2010). If even 1 of the 19 birds found to be viremic for WNV in the recent North American field survey (Dusek et al. 2009) could have been fitted with a radiotransmitter, our understanding of the potential for arbovirus-infected birds to move in the environment would advance enormously.

The difficulties with maintaining exclusively insectivorous migratory birds in the laboratory for experimental infection studies (Reisen et al. 2010) need to be addressed so that we can begin to understand how different arboviruses affect these species (e.g., onset, strength, and duration of viremias; antibody production and persistence; mortality rates) in the same way we have done for resident birds, especially if these species allocate investment in the immune system differently (Møller and Erritzøe 1998). Because of the laborious nature of sampling birds in the field and the infrequency with which birds with arbovirus infections are detected despite large sample sizes (Table 1), molecular analyses of vector blood meals that can identify the host species being fed upon (e.g., Ngo and Kramer 2003, Kilpatrick et al. 2006c, Molaei et al. 2008, Watts et al. 2009) can be used to determine the likelihood that resident versus migratory species are potentially exposed to arboviruses. Combined with field surveys of bird species abundance taken at the same time that blooded mosquitoes are collected, we now have the molecular tools to investigate whether the reduced seroprevalence of migratory species for particular arboviruses (Day and Stark 1999; Reisen et al. 2003c, 2010) reflects these species' lower exposure to viruses as a result of mosquito feeding preferences, higher mortality, or more rapid degradation of antibodies.

A significant gap in our understanding of arbovirus transmission in general is knowing how likely a given bird is to be fed upon by transmission-competent vectors when these vectors are present in the same habitat as the bird. A number of studies on mosquito biting rates of birds in captivity have been conducted (Blackmore and Dow 1958; Edman et al. 1972; Scott et al. 1988, 1990; Hodgson et al. 2001), but we know almost nothing about how often birds of different ages, sexes, or residence status are fed upon by mosquitoes in the wild (Griffing et al. 2007). An exception was a field study of American Robins that used infrared cameras, showing that some adult birds attracted as many as 500 *Culex* landings per night while others had <50 (Griffing et al. 2007). Although many of these landings may not have resulted in a blood meal, same-age birds of the same species likely differed substantially in their rate of exposure to vectors that might transmit arboviruses. Similar work on other bird species is needed to determine whether certain behavioral, ecological, or morphological characteristics make an individual more or less likely to attract mosquitoes or to engage in anti-mosquito defensive behavior, and, if so, whether these traits vary with avian migratory propensity. Some data indicate that certain bird species are more likely to engage in effective anti-mosquito defensive behavior, resulting in interrupted feeding by mosquitoes and in the mosquitoes' biting of multiple host individuals; in such cases, arbovirus transmission among more susceptible birds may be enhanced, especially in roosts (Hodgson et al. 2001).

We need to evaluate how bird transport of arboviruses may directly affect the evolution of virulence in some viruses: are more asymptomatic strains that affect the movement of birds less by not making them as ill (while still causing high enough viremia to infect vectors) more likely to be transported over wide areas and thus potentially outcompete the more virulent virus subtypes (Komar and Clark 2006, Altizer et al. 2011)? More generally, because of transmission–virulence tradeoffs, the relatively benign strains may be the ones most likely to exploit opportunities to infect (and be transported by) multiple host species in the wild, in contrast to the more highly pathogenic viruses that are better adapted to transmission among closely spaced individuals of single host species (e.g., high-pathogenic avian influenza viruses in domestic poultry; Lebarbenchon et al. 2010).

Future work on local transport of arboviruses should perhaps focus on highly social bird species that represent dense resource patches for blood-feeding arthropod vectors, such as American Crows (Reisen et al. 2006a, Nielsen and Reisen 2007), ardeids (Bowen and McLean 1977), American Robins and European Starlings (Komar 1997, Hodgson et al. 2001), or swallows (Brown and Sethi 2002), and that are less constrained spatially to defended territories. More generally, the importance of bird social structure and the extent to which aggregations of hosts are important in transmission and persistence of arboviruses are largely unexplored. Some theoretical models suggest that as birds aggregate into colonies or roosts, amplification and potential spread of viruses such as WNV could increase (Shaman 2007). An explicit focus on bird sociospatial distribution may be especially important if some arboviruses have a greater likelihood of direct bird-to-bird transmission than we generally assume (Holden 1955b, McLean et al. 2001, Banet-Noach et al. 2003, N. Komar et al. 2003, Dawson et al. 2007, Hartemink et al. 2007, Huyvaert et al. 2008). Direct transmission frees birds to potentially disperse viruses more effectively because it removes the restrictive requirements of having appropriate vectors feed on them at certain times and places in order for transmission to occur. Additional work is clearly needed to determine the extent to which arboviruses can be successfully dispersed via ingestion of carcasses or by oral or cloacal shedding. Finally, more attention should be given to bird species that have closely coevolved with their respective arboviruses, such as EEEV in wading birds in the southeastern United States (Kissling et al. 1954, McLean et al. 1995) and BCRV in Cliff Swallows (Brown et al. 2007, 2008; O'Brien et al. 2011). These species seem to be less negatively affected, either because of evolved immunity to their associated viruses or because the viruses have attenuated over time to become less virulent (Ewald 1994, Lebarbenchon et al. 2010).

The cautions we voice here about inferring bird involvement in arbovirus transport have also been raised about the frequent assumption that wild birds are responsible for intercontinental transport of avian influenza viruses (Anonymous 2006, Melville and Shortridge 2006, Yasue et al. 2006, Gauthier-Clerc et al. 2007, Weber and Stilianakis 2007, Winker et al. 2007, Stoops et al. 2009). Although considerable effort has been directed at sampling large numbers of wild birds across multiple continents for the highly pathogenic H5N1 strain of avian influenza, to date there is little convincing empirical evidence that infected and asymptomatic birds ever carry this influenza virus (or less pathogenic strains) along established long-distance migratory routes (Feare and Yasue 2006, Yasue et al. 2006, Feare 2007, Gauthier-Clerc et al. 2007, Krauss et al. 2007, van Gils et al. 2007, Pearce et al. 2009). The costs to an infected bird of mounting an immune response to avian influenza viruses (and possibly to viruses more generally) may curtail or greatly reduce its long-distance movement during the period of infection (Weber and Stilianakis 2007).

The tiny fraction of birds reported to be viremic in field surveys (Table 1) may be enough to result in meaningful arbovirus transport that can serve to mix virus haplotypes over different geographic areas, reintroduce virus to regions where transmission is interrupted in winter, or, in rare cases, result in virus colonization of new locales. Unfortunately, we cannot know how plausible these outcomes are without better information on the frequency with which birds that are actually moving carry transmissible arboviruses. Considerable work is needed to substantiate the assumption that arboviruses are routinely transported by wild birds over even short distances. Premature conclusions about the role of birds as pathogen reservoirs or transport agents could have an unintended but serious negative effect on conservation of many migratory species throughout the world (Weber and Stilianakis 2007) and cause public health resources to be diverted into what might ultimately prove to be ineffective ways to predict or prevent disease spread.

Acknowledgments

For helping shape our thoughts on birds and arboviruses and/or for comments on the manuscript, we thank K. Huyvaert, N. Komar, K. Miller, A. Moore, A. Padhi, and M. Pfeffer. We are grateful especially to W. Reisen, who made numerous helpful suggestions on two drafts of the manuscript. Our work on viruses and birds has been supported by grants (to C.R.B.) from the National Institutes of Health (AI057569) and the National Science Foundation (DEB-0075199, DEB-0514824), and (to V.A.O.) from the American Ornithologists' Union.

Literature Cited

Able, K. P. 1972. Fall migration in coastal Louisiana and the evolution of migration patterns in the Gulf region. Wilson Bulletin 84:231–242.

Aitken, T. H. G., W. G. Downs, L. Spence, and A. H. Jonkers. 1964. St. Louis encephalitis virus isolations in Trinidad, West Indies, 1953–1962. American Journal of Tropical Medicine and Hygiene 13:450–451.

Alerstam, T., and A. Lindstrom. 1990. Optimal bird migration: The relative importance of time, energy, and safety. Pages 331–351 in Bird Migration: Physiology and Ecophysiology (E. Gwinner, Ed.). Springer-Verlag, Berlin.

Altizer, S., R. Bartel, and B. A. Han. 2011. Animal migration and infectious disease risk. Science 331:296–302.

Anderson, J. F., and A. J. Main. 2006. Importance of vertical and horizontal transmission of West Nile virus by Culex pipiens in the northeastern United States. Journal of Infectious Diseases 194:1577–1579.

Andreadis, T. G., J. F. Anderson, C. R. Vossbrinck, and A. J. Main. 2004. Epidemiology of West Nile virus in Connecticut: A five-year analysis of mosquito data 1999–2003. Vector-Borne and Zoonotic Diseases 4:360–378.

Anonymous. 2006. Avian influenza goes global, but don't blame the birds. Lancet Infectious Diseases 6:185.

ARMSTRONG, P. M., AND T. G. ANDREADIS. 2006. A new genetic variant of La Crosse virus (Bunyaviridae) isolated from New England. American Journal of Tropical Medicine and Hygiene 75:491–496.

BÄCHLER, E., S. HAHN, M. SCHAUB, R. ARLETTAZ, L. JENNI, J. W. FOX, V. AFANASYEV, AND F. LIECHTI. 2010. Year-round tracking of small trans-Sahara migrants using light-level geolocators. PLoS One, e9566.

BAILEY, C. L., B. F. ELDRIDGE, D. E. HAYES, D. M. WATTS, R. F. TAMMARIELLO, AND J. M. DALRYMPLE. 1978. Isolation of St. Louis encephalitis virus from overwintering *Culex pipiens* mosquitoes. Science 199: 1346–1349.

BALANÇA, G., N. GAIDET, G. SAVINI, B. VOLLOT, A. FOUCART, P. REITER, A. BOUTONNIER, R. LELLI, AND F. MONICAT. 2009. Low West Nile virus circulation in wild birds in an area of recurring outbreaks in southern France. Vector-Borne and Zoonotic Diseases 9:737–741.

BALDUCCI, M., P. VERANI, M. C. LOPES, AND B. GREGORIG. 1973. Isolation in Italy of Bahig and Matruh viruses (Tete group) from migratory birds. Annals of Microbiology (Institute Pasteur) 124B:231–237.

BANET-NOACH, C., L. SIMANOV, AND M. MALKINSON. 2003. Direct (non-vector) transmission of West Nile virus in geese. Avian Pathology 32:489–494.

BARRERA, J. DEL C., AND G. J. LETCHWORTH. 1996. Persistence of vesicular stomatitis virus New Jersey RNA in convalescent hamsters. Virology 219:453–464.

BAST, T. F., E. WHITNEY, AND J. L. BENACH. 1973. Considerations on the ecology of several arboviruses in eastern Long Island. American Journal of Tropical Medicine and Hygiene 22:109–115.

BAUCHINGER, U., AND M. KLAASSEN. 2005. Longer days in spring than in autumn accelerate migration speed of passerine birds. Journal of Avian Biology 36:3–5.

BAYLY, N. J. 2006. Optimality in avian migratory fuelling behaviour: A study of a trans-Saharan migrant. Animal Behaviour 71:173–182.

BENGIS, R. G., F. A. LEIGHTON, J. R. FISHER, M. ARTOIS, T. MÖRNER, AND C. M. TATE. 2004. The role of wildlife in emerging and re-emerging zoonoses. Revue Scientifique et Technique 23:497–511.

BERGE, T. O., ED. 1975. International Catalogue of Arboviruses Including Certain Other Viruses of Vertebrates, 2nd ed. No. (CDC) 75-8301. Public Health Service, U.S. Department of Health, Education, and Welfare, Washington, D.C.

BERGE, T. O., R. W. CHAMBERLAIN, R. E. SHOPE, AND T. H. WORK. 1971. Catalogue of arthropod-borne and selected vertebrate viruses of the world. American Journal of Tropical Medicine and Hygiene 20:1018–1050.

BERTHET, F.-X., H. G. ZELLER, M.-T. DROUET, J. RAUZIER, J.-P. DIGOUTTE, AND V. DEUBEL. 1997. Extensive nucleotide changes and deletions within the envelope glycoprotein gene of Euro-African West Nile viruses. Journal of General Virology 78:2293–2297.

BERTOLOTTI, L., U. KITRON, AND T. L. GOLDBERG. 2007. Diversity and evolution of West Nile virus in Illinois and the United States, 2002–2005. Virology 360:143–149.

BERTOLOTTI, L., U. D. KITRON, E. D. WALKER, M. O. RUIZ, J. D. BRAWN, S. R. LOSS, G. L. HAMER, AND T. L. GOLDBERG. 2008. Fine-scale genetic variation and evolution of West Nile virus in a transmission "hot spot" in suburban Chicago, USA. Virology 374:381–389.

BIEBACH, H., W. FRIEDRICH, AND G. HEINE. 1986. Interaction of body mass, fat, foraging and stopover period in trans-Sahara migrating passerine birds. Oecologia 69:370–379.

BJÖERSDORFF, A., S. BERGSTRÖM, R. F. MASSUNG, P. D. HAEMIG, AND B. OLSEN. 2001. *Ehrlichia*-infected ticks on migrating birds. Emerging Infectious Diseases 7:877–879.

BLACKMORE, J. S., AND R. P. DOW. 1958. Differential feeding of *Culex tarsalis* on nestling and adult birds. Mosquito News 18:15–17.

BLONDEL, J. 1972. Bird migrations in western Europe. Pages 26–32 *in* Transcontinental Connections of Migratory Birds and Their Role in the Distribution of Arboviruses. Nauka, Academy of Sciences of the USSR, Siberian Branch, Biological Institute, Novosibirsk, Siberia (USSR).

BOND, J. O., W. L. JENNINGS, AND G. E. WOOLFENDEN. 1972. Arbovirus investigations among migratory birds on the Florida peninsula, USA. Pages 205–206 *in* Transcontinental Connections of Migratory Birds and their Role in the Distribution of Arboviruses. Nauka, Academy of Sciences of the USSR, Siberian Branch, Biological Institute, Novosibirsk, Siberia (USSR).

BOWEN, G. S., AND R. G. MCLEAN. 1977. Experimental infection of birds with epidemic Venezuelan encephalitis virus. American Journal of Tropical Medicine and Hygiene 26:808–814.

BOWEN, G. S., T. P. MONATH, G. E. KEMP, J. H. KERSCHNER, AND L. J. KIRK. 1980. Geographic variation among St. Louis encephalitis virus strains in the viremic responses of avian hosts. American Journal of Tropical Medicine and Hygiene 29:1411–1419.

BOYLE, D. B., R. W. DICKERMAN, AND I. D. MARSHALL. 1983. Primary viremia responses of herons to experimental infection with Murray Valley encephalitis, Kunjin, and Japanese encephalitis viruses. Australian Journal of Experimental Biology and Medical Science 61:655–664.

BRAULT, A. C., M. V. ARMIJOS, S. WHEELER, S. WRIGHT, Y. FANG, S. LANGEVIN, AND W. K. REISEN. 2009. Stone Lakes virus (family Togaviridae, genus *Alphavirus*), a variant of Fort Morgan virus isolated from swallow bugs (Hemiptera: Cimicidae) west of the Continental Divide. Journal of Medical Entomology 46: 1203–1209.

BRAULT, A. C., C. Y.-H. HUANG, S. A. LANGEVIN, R. M. KINNEY, R. A. BOWEN, W. N. RAMEY, N. A. PANELLA,

E. C. Holmes, A. M. Powers, and B. R. Miller. 2007. A single positively selected West Nile viral mutation confers increased virogenesis in American Crows. Nature Genetics 39:1162–1166.

Brault, A. C., A. M. Powers, C. L. Villarreal, R. N. Lopez, M. F. Cachon, L. F. L. Gutierrez, W. Kang, R. B. Tesh, R. E. Shope, and S. C. Weaver. 1999. Genetic and antigenic diversity among eastern equine encephalitis viruses from North, Central, and South America. American Journal of Tropical Medicine and Hygiene 61:579–586.

Brown, C. R. 1978. Post-fledging behavior of Purple Martins. Wilson Bulletin 90:376–385.

Brown, C. R., and M. B. Brown. 1996. Coloniality in the Cliff Swallow: The Effect of Group Size on Social Behavior. University of Chicago Press, Chicago, Illinois.

Brown, C. R., and M. B. Brown. 2000. Weather-mediated natural selection on arrival time in Cliff Swallows (*Petrochelidon pyrrhonota*). Behavioral Ecology and Sociobiology 47:339–345.

Brown, C. R., and M. B. Brown. 2004. Empirical measurement of parasite transmission between groups in a colonial bird. Ecology 85:1619–1626.

Brown, C. R., and M. B. Brown. 2005. Between-group transmission dynamics of the swallow bug, *Oeciacus vicarius*. Journal of Vector Ecology 30:137–143.

Brown, C. R., M. B. Brown, A. T. Moore, and N. Komar. 2007. Bird movement predicts Buggy Creek virus infection in insect vectors. Vector-Borne and Zoonotic Diseases 7:304–314.

Brown, C. R., M. B. Brown, A. Padhi, J. E. Foster, A. T. Moore, M. Pfeffer, and N. Komar. 2008. Host and vector movement affects genetic diversity and spatial structure of Buggy Creek virus (Togaviridae). Molecular Ecology 17:2164–2173.

Brown, C. R., N. Komar, S. B. Quick, R. A. Sethi, N. A. Panella, M. B. Brown, and M. Pfeffer. 2001. Arbovirus infection increases with group size. Proceedings of the Royal Society of London, Series B 268:1833–1840.

Brown, C. R., A. T. Moore, S. A. Knutie, and N. Komar. 2009a. Overwintering of infectious Buggy Creek virus (Togaviridae, Alphavirus) in *Oeciacus vicarius* (Hemiptera: Cimicidae) in North Dakota. Journal of Medical Entomology 46:391–394.

Brown, C. R., A. T. Moore, G. R. Young, and N. Komar. 2010a. Persistence of Buggy Creek virus (Togaviridae, *Alphavirus*) for two years in unfed swallow bugs (Hemiptera: Cimicidae: *Oeciacus vicarius*). Journal of Medical Entomology 47:436–441.

Brown, C. R., A. T. Moore, G. R. Young, A. Padhi, and N. Komar. 2009b. Isolation of Buggy Creek virus (Togaviridae, *Alphavirus*) from field-collected eggs of *Oeciacus vicarius* (Hemiptera: Cimicidae). Journal of Medical Entomology 46:375–379.

Brown, C. R., A. Padhi, A. T. Moore, M. B. Brown, J. E. Foster, M. Pfeffer, V. A. O'Brien, and N. Komar. 2009c. Ecological divergence of two sympatric lineages of Buggy Creek virus, an arbovirus associated with birds. Ecology 90:3168–3179.

Brown, C. R., and R. A. Sethi. 2002. Mosquito abundance is correlated with Cliff Swallow (*Petrochelidon pyrrhonota*) colony size. Journal of Medical Entomology 39:115–120.

Brown, C. R., S. A. Strickler, A. T. Moore, S. A. Knutie, A. Padhi, M. B. Brown, G. R. Young, V. A. O'Brien, J. E. Foster, and N. Komar. 2010b. Winter ecology of Buggy Creek virus (Togaviridae, *Alphavirus*) in the central Great Plains. Vector Borne and Zoonotic Diseases 10:355–363.

Brust, R. A. 1980. Dispersal behavior of adult *Aedes stricticus* and *Aedes vexans* (Diptera: Culicidae) in Manitoba. Canadian Entomologist 112:31–42.

Buckley, A., A. Dawson, S. R. Moss, S. A. Hinsley, P. E. Bellamy, and E. A. Gould. 2003. Serological evidence of West Nile virus, Usutu virus, and Sindbis virus infection of birds in the UK. Journal of General Virology 84:2807–2817.

Buescher, E. L., W. F. Scherer, H. E. McClure, J. T. Moyer, M. Z. Rosenberg, M. Yoshii, and Y. Okada. 1959. Ecologic studies of Japanese encephalitis virus in Japan. IV. Avian infection. American Journal of Tropical Medicine and Hygiene 8:678–688.

Burke, D. S., and C. J. Leake. 1988. Japanese encephalitis. Pages 63–92 *in* The Arboviruses: Epidemiology and Ecology, vol. 3 (T. P. Monath, Ed.). CRC Press, Boca Raton, Florida.

Burkett-Cadena, N. D., R. A. Ligon, M. Liu, H. K. Hassan, G. E. Hill, M. D. Eubanks, and T. R. Unnasch. 2010. Vector–host interactions in avian nests: Do mosquitoes prefer nestlings over adults? American Journal of Tropical Medicine and Hygiene 83:395–399.

Burt, F. J., A. A. Grobbelaar, P. A. Leman, F. S. Anthony, G. V. F. Gibson, and R. Swanepoel. 2002. Phylogenetic relationships of southern Africa West Nile virus isolates. Emerging Infectious Diseases 8:820–826.

Cabe, P. R. 1993. European Starling (*Sturnus vulgaris*). *In* The Birds of North America, no. 48 (A. Poole and F. Gill, Eds.). Academy of Natural Sciences, Philadelphia, and American Ornithologists' Union, Washington, D.C.

Caffrey, C., S. C. R. Smith, and J. T. Weston. 2005. West Nile virus devastates an American Crow population. Condor 107:128–132.

Calisher, C. H. 1994. Medically important arboviruses of the United States and Canada. Clinical Microbiology Reviews 7:89–116.

Calisher, C. H., E. V. Gutiérrez, K. S. C. Maness, and R. D. Lord. 1974. Isolation of Mayaro virus from a migrating bird captured in Louisiana in 1967. Bulletin of the Pan American Health Organization 8:243–248.

Calisher, C. H., N. Karabatsos, J. S. Lazuick, T. P. Monath, and K. L. Wolff. 1988. Reevaluation of the western equine encephalitis antigenic complex of alphaviruses (family Togaviridae) as determined by neutralization tests. American Journal of Tropical Medicine and Hygiene 38:447–452.

CALISHER, C. H., K. S. C. MANESS, R. D. LORD, AND P. H. COLEMAN. 1971. Identification of two South American strains of eastern equine encephalomyelitis virus from migrant birds captured on the Mississippi delta. American Journal of Epidemiology 94:172–178.

CHANDLER, L. J., R. PARSONS, AND Y. RANDLE. 2001. Multiple genotypes of St. Louis encephalitis virus (Flaviviridae: *Flavivirus*) circulate in Harris County, Texas. American Journal of Tropical Medicine and Hygiene 64:12–19.

CHARREL, R. N., A. C. BRAULT, P. GALLIAN, J.-L. LEMASSON, B. MURGUE, S. MURRI, B. PASTORINO, H. ZELLER, R. DE CHESSE, P. DE MICCO, AND X. DE LAMBALLERIE. 2003. Evolutionary relationship between Old World West Nile virus strains: Evidence for viral gene flow between Africa, the Middle East, and Europe. Virology 315:381–388.

CHAVEZ-RAMIREZ, F., G. P. VOSE, AND A. TENNANT. 1994. Spring and fall migration of Peregrine Falcons from Padre Island, Texas. Wilson Bulletin 106:138–145.

CHEVALIER, V., P. REYNAUD, T. LEFRANCOIS, B. DURAND, F. BAILLON, G. BALANÇA, N. GAIDET, B. MONDET, AND R. LANCELOT. 2009. Predicting West Nile virus seroprevalence in wild birds in Senegal. Vector-Borne and Zoonotic Diseases 9:589–596.

CHILTON, G., M. C. BAKER, C. D. BARRENTINE, AND M. A. CUNNINGHAM. 1995. White-crowned Sparrow (*Zonotrichia leucophrys*). *In* The Birds of North America, no. 183 (A. Poole and F. Gill, Eds.). Academy of Natural Sciences, Philadelphia, and American Ornithologists' Union, Washington, D.C.

CHOI, S., AND S. C. JIANG. 2005. Real-time PCR quantification of human adenoviruses in urban rivers indicates genome prevalence but low infectivity. Applied and Environmental Microbiology 71:7426–7433.

CILNIS, M. J., W. KANG, AND S. C. WEAVER. 1996. Genetic conservation of Highlands J viruses. Virology 218:343–351.

CIMPRICH, D. A., AND F. R. MOORE. 1995. Gray Catbird (*Dumetella carolinensis*). *In* The Birds of North America, no. 167 (A. Poole and F. Gill, Eds.). Academy of Natural Sciences, Philadelphia, and American Ornithologists' Union, Washington, D.C.

COCHRAN, W. W. 1987. Orientation and other migratory behaviors of a Swainson's Thrush followed for 1500 km. Animal Behaviour 35:927–929.

COCHRAN, W. W., AND M. WIKELSKI. 2005. Individual migratory tactics of New World *Catharus* thrushes: Current knowledge and future tracking options from space. Pages 274–289 *in* Birds of Two Worlds: The Ecology and Evolution of Migration (R. Greenberg and P. P. Marra, Eds.). Johns Hopkins University Press, Baltimore, Maryland.

COCKBURN, T. A., C. A. SOOTER, AND A. D. LANGMUIR. 1957. Ecology of western equine and St. Louis encephalitis viruses: A summary of field investigations in Weld County, Colorado, 1949 to 1953. American Journal of Hygiene 65:130–146.

CONVERSE, J. D., H. HOOGSTRAAL, M. I. MOUSSA, M. STEK, JR., AND M. N. KAISER. 1974. Bahig virus (Tete group) in naturally- and transovarially-infected *Hyalomma marginatum* ticks from Egypt and Italy. Archives of Virology 46:29–35.

CRANS, W. J., D. F. CACCAMISE, AND J. R. MCNELLY. 1994. Eastern equine encephalomyelitis virus in relation to the avian community of a coastal cedar swamp. Journal of Medical Entomology 31:711–728.

CUPP, E. W., K. J. TENNESSEN, W. K. OLDLAND, H. K. HASSAN, G. E. HILL, C. R. KATHOLI, AND T. R. UNNASCH. 2004. Mosquito and arbovirus activity during 1997–2002 in a wetland in northeastern Mississippi. Journal of Medical Entomology 41:495–501.

DALRYMPLE, J. M., O. P. YOUNG, B. F. ELDRIDGE, AND P. K. RUSSELL. 1972. Ecology of arboviruses in a Maryland freshwater swamp. American Journal of Epidemiology 96:129–140.

DAVIS, C. T., D. W. C. BEASLEY, H. GUZMAN, P. RAJ, M. D'ANTON, R. J. NOVAK, T. R. UNNASCH, R. B. TESH, AND A. D. T. BARRETT. 2003. Genetic variation among temporally and geographically distinct West Nile virus isolates, United States, 2001, 2002. Emerging Infectious Diseases 9:1423–1429.

DAVIS, C. T., G. D. EBEL, R. S. LANCIOTTI, A. C. BRAULT, H. GUZMAN, M. SIIRIN, A. LAMBERT, R. E. PARSONS, D. W. C. BEASLEY, R. J. NOVAK, AND OTHERS. 2005. Phylogenetic analysis of North American West Nile virus isolates, 2001–2004: Evidence for the emergence of a dominant genotype. Virology 342:252–265.

DAWSON, J. R., W. B. STONE, G. D. EBEL, D. S. YOUNG, D. S. GALINSKI, J. P. PENSABENE, M. A. FRANKE, M. EIDSON, AND L. D. KRAMER. 2007. Crow deaths caused by West Nile virus during winter. Emerging Infectious Diseases 13:1912–1914.

DAY, J. F. 2001. Predicting St. Louis encephalitis virus epidemics: Lessons from recent, and not so recent, outbreaks. Annual Review of Entomology 46:111–138.

DAY, J. F., AND L. M. STARK. 1999. Avian serology in a St. Louis encephalitis epicenter before, during, and after a widespread epidemic in south Florida, USA. Journal of Medical Entomology 36:614–624.

DE THOISY, B., J. GARDON, R. A. SALAS, J. MORVAN, AND M. KAZANJI. 2003. Mayaro virus in wild mammals, French Guiana. Emerging Infectious Diseases 9:1326–1329.

DELINGAT, J., F. BAIRLEIN, AND A. HEDENSTRÖM. 2008. Obligatory barrier crossing and adaptive fuel management in migratory birds: The case of the Atlantic crossing in Northern Wheatears (*Oenanthe oenanthe*). Behavioral Ecology and Sociobiology 62:1069–1078.

DIAZ, L. A., N. KOMAR, A. VISINTIN, M. J. D. JURI, M. STEIN, R. L. ALLENDE, L. SPINSANTI, B. KONIGHEIM, J. AGUILAR, M. LAURITO, AND OTHERS. 2008. West Nile virus in birds, Argentina. Emerging Infectious Diseases 14:689–691.

DICKERMAN, R. W., M. S. MARTIN, AND E. A. DIPAOLA. 1980. Studies of Venezuelan encephalitis in migrating birds in relation to possible transport of virus from South to Central America. American Journal of Tropical Medicine and Hygiene 29:269–276.

DOWNS, W. G., T. H. G. AITKEN, AND L. SPENCE. 1959. Eastern equine encephalitis virus isolated from *Culex nigripalpus* in Trinidad. Science 139:1471.

DOWNS, W. G., C. R. ANDERSON, AND J. CASALS. 1957. The isolation of St. Louis virus from a nestling bird in Trinidad, British West Indies. American Journal of Tropical Medicine and Hygiene 6:693–696.

DRAKE, V. A., AND R. A. FARROW. 1988. The influence of atmospheric structure and motions on insect migration. Annual Review of Entomology 33:183–210.

DUPUIS, A. P., II, P. P. MARRA, AND L. D. KRAMER. 2003. Serologic evidence of West Nile virus transmission, Jamaica, West Indies. Emerging Infectious Diseases 9:860–863.

DUSEK, R. J., R. G. MCLEAN, L. D. KRAMER, S. R. UBICO, A. P. DUPUIS II, G. D. EBEL, AND S. C. GUPTILL. 2009. Prevalence of West Nile virus in migratory birds during spring and fall migration. American Journal of Tropical Medicine and Hygiene 81:1151–1158.

EDMAN, J. D., L. A. WEBBER, AND H. W. KALE II. 1972. Effect of mosquito density on the interrelationship of host behavior and mosquito feeding success. American Journal of Tropical Medicine and Hygiene 21:487–491.

EDMAN, J. D., L. A. WEBBER, AND A. A. SCHMID. 1974. Effect of host defenses on the feeding pattern of *Culex nigripalpus* when offered a choice of blood sources. Journal of Parasitology 60:874–883.

EIDSON, M., L. KRAMER, W. STONE, Y. HAGIWARA, K. SCHMIT, AND THE NEW YORK STATE WEST NILE VIRUS AVIAN SURVEILLANCE TEAM. 2001. Dead bird surveillance as an early warning system for West Nile virus. Emerging Infectious Diseases 7:631–635.

EMORD, D. E., AND C. D. MORRIS. 1984. Epizootiology of eastern equine encephalomyelitis virus in upstate New York, USA. VI. Antibody prevalence in wild birds during an interepizootic period. Journal of Medical Entomology 21:395–404.

ENSERINK, M. 2009. Infection study worries farmers, bird lovers. Science 323:324.

ERLANGER, T. E., S. WEISS, J. KEISER, J. UTZINGER, AND K. WIEDENMAYER. 2009. Past, present, and future of Japanese encephalitis. Emerging Infectious Diseases 15:1–7.

ERNEK, E., O. KOŽUCH, M. GREŠÍKOVÁ, J. NOSEK, AND M. SEKEYOVÁ. 1973. Isolation of Sindbis virus from the Reed Warbler (*Acrocephalus scirpaceus*) in Slovakia. Acta Virologica 17:359–361.

ERNEK, E., O. KOŽUCH, M. LICHARD, AND J. NOSEK. 1968. The role of birds in the circulation of tick-borne encephalitis virus in the Tribeč region. Acta Virologica 12:468–470.

ERNEK, E., O. KOŽUCH, J. NOSEK, J. TEPLAN, AND C. FOLK. 1977. Arboviruses in birds captured in Slovakia. Journal of Hygiene, Epidemiology, Microbiology and Immunology 21:353–359.

ESTRADA-FRANCO, J. G., R. NAVARRO-LOPEZ, D. W. C. BEASLEY, L. COFFEY, A.-S. CARRARA, A. T. DA ROSA, T. CLEMENTS, E. WANG, G. V. LUDWIG, A. C. CORTES, AND OTHERS. 2003. West Nile virus in Mexico: Evidence of widespread circulation since July 2002. Emerging Infectious Diseases 9:1604–1607.

EVANS, B. E., AND T. A. SORDAHL. 2009. Factors influencing perch selection by communally roosting Turkey Vultures. Journal of Field Ornithology 80:364–372.

EWALD, P. W. 1994. Evolution of Infectious Disease. Oxford University Press, Oxford, United Kingdom.

FAIRLEY, T. L., T. M. RENAUD, AND J. E. CONN. 2000. Effects of local geographic barriers and latitude on population structure in *Anopheles punctipennis* (Diptera: Culicidae). Journal of Medical Entomology 37:754–760.

FEARE, C. J. 2007. The role of wild birds in the spread of HPAI H5N1. Avian Diseases 51:440–447.

FEARE, C. J., AND M. YASUE. 2006. Asymptomatic infection with highly pathogenic avian influenza H5N1 in wild birds: How sound is the evidence? Virology Journal 3:96.

FIGUEROLA, J., M. A. JIMENEZ-CLAVERO, G. ROJO, C. GOMEZ-TAYLOR, AND R. SORIGUER. 2007. Prevalence of West Nile virus neutralizing antibodies in colonial aquatic birds in southern Spain. Avian Pathology 36:209–212.

FORRESTER, N. L., J. L. KENNEY, E. DEARDORFF, E. WANG, AND S. C. WEAVER. 2008. Western equine encephalitis submergence: Lack of evidence for a decline in virus virulence. Virology 380:170–172.

GARVIN, M. C., K. A. TARVIN, L. M. STARK, G. E. WOOLFENDEN, J. W. FITZPATRICK, AND J. F. DAY. 2004. Arboviral infection in two species of wild jays (Aves: Corvidae): Evidence for population impacts. Journal of Medical Entomology 41:215–225.

GAUTHIER-CLERC, M., C. LEBARBENCHON, AND F. THOMAS. 2007. Recent expansion of highly pathogenic avian influenza H5N1: A critical review. Ibis 149:202–214.

GEBHARDT, L. P., AND D. W. HILL. 1960. Overwintering of western equine encephalitis virus. Proceedings of the Society of Experimental Biology and Medicine 104:695–698.

GEORGOPOULOU, I., AND V. TSIOURIS. 2008. The potential role of migratory birds in the transmission of zoonoses. Veterinaria Italiana 44:671–677.

GILBERT, M., X. XIAO, J. DOMENECH, J. LUBROTH, V. MARTIN, AND J. SLINGENBERGH. 2006. Anatidae migration in the Western Palearctic and spread of highly pathogenic avian influenza H5N1 virus. Emerging Infectious Diseases 12:1650–1656.

GILL, R. E., JR., T. PIERSMA, G. HUFFORD, R. SERVRANCKX, AND A. RIEGEN. 2005. Crossing the ultimate ecological barrier: Evidence for an 11000-km-long nonstop flight from Alaska to New Zealand and Australia by Bar-tailed Godwits. Condor 107:1–20.

GODDARD, L. B., A. E. ROTH, W. K. REISEN, AND T. W. SCOTT. 2002. Vector competence of California mosquitoes for West Nile virus. Emerging Infectious Diseases 8:1385–1391.

GODDARD, L. B., A. E. ROTH, W. K. REISEN, AND T. W. SCOTT. 2003. Vertical transmission of West Nile virus by three California *Culex* (Diptera: Culicidae) species. Journal of Medical Entomology 40:743–746.

GOLDBERG, T. L., T. K. ANDERSON, AND G. L. HAMER. 2010. West Nile virus may have hitched a ride across the western United States on *Culex tarsalis* mosquitoes. Molecular Ecology 19:1518–1519.

GOTTDENKER, N. L., E. W. HOWERTH, AND D. G. MEAD. 2003. Natural infection of a Great Egret (*Casmerodius albus*) with eastern equine encephalitis virus. Journal of Wildlife Diseases 39:702–706.

GOULD, E. A., X. DE LAMBALLERIE, P. M. A. ZANOTTO, AND E. C. HOLMES. 2001. Evolution, epidemiology, and dispersal of flaviviruses revealed by molecular phylogenies. Advances in Virus Research 57:71–103.

GOULD, E. A., AND S. HIGGS. 2009. Impact of climate change and other factors on emerging arbovirus diseases. Transactions of the Royal Society of Tropical Medicine and Hygiene 103:109–121.

GOULD, E. A., S. R. MOSS, AND S. L. TURNER. 2004. Evolution and dispersal of encephalitic flaviviruses. Pages 65–84 *in* Emergence and Control of Zoonotic Viral Encephalitides (C. H. Calisher and D. E. Griffin, Eds.). Springer, New York.

GRIFFING, S. M., A. M. KILPATRICK, L. CLARK, AND P. P. MARRA. 2007. Mosquito landing rates on nestling American Robins (*Turdus migratorius*). Vector-Borne and Zoonotic Diseases 7:437–443.

GRINEV, A., S. DANIEL, S. STRAMER, S. ROSSMANN, S. CAGLIOTI, AND M. RIOS. 2008. Genetic variability of West Nile virus in US blood donors, 2002–2005. Emerging Infectious Diseases 14:436–444.

GRUWELL, J. A., C. L. FOGARTY, S. G. BENNETT, G. L. CHALLET, K. S. VANDERPOOL, M. JOZAN, AND J. P. WEBB, JR. 2000. Role of peridomestic birds in the transmission of St. Louis encephalitis virus in southern California. Journal of Wildlife Diseases 36:13–34.

GUBLER, D. J. 2002. The global emergence/resurgence of arboviral diseases as public health problems. Archives of Medical Research 33:330–342.

GYLFE, A., S. BERGSTRÖM, J. LUNDSTRÖM, AND B. OLSEN. 2000. Reactivation of *Borrelia* infection in birds. Nature 403:724–725.

HAMER, G. L., E. D. WALKER, J. D. BRAWN, S. R. LOSS, M. O. RUIZ, T. L. GOLDBERG, A. M. SCHOTTHOEFER, W. M. BROWN, E. WHEELER, AND U. D. KITRON. 2008. Rapid amplification of West Nile virus: The role of hatch-year birds. Vector-Borne and Zoonotic Diseases 8:57–67.

HAMMON, W. M., W. C. REEVES, AND G. E. SATHER. 1951. Western equine and St. Louis encephalitis viruses in the blood of experimentally infected wild birds and epidemiological implications of findings. Journal of Immunology 67:357–367.

HANNA, J. N., S. A. RITCHIE, D. A. PHILLIPS, J. SHIELD, M. C. BAILEY, J. S. MACKENZIE, M. POIDINGER, B. J. MCCALL, AND P. J. MILLS. 1996. An outbreak of Japanese encephalitis in the Torres Strait, Australia, 1995. Medical Journal of Australia 165:256–260.

HANNOUN, C., B. CORNIOU, AND J. MOUCHET. 1972. Role of migrating birds in arbovirus transfer between Africa and Europe. Pages 167–172 *in* Transcontinental Connections of Migratory Birds and Their Role in the Distribution of Arboviruses. Nauka, Academy of Sciences of the USSR, Siberian Branch, Biological Institute, Novosibirsk, Siberia (USSR).

HARDY, J. L., E. J. HOUK, L. D. KRAMER, AND W. C. REEVES. 1983. Intrinsic factors affecting vector competence of mosquitoes for arboviruses. Annual Review of Entomology 28:229–262.

HARDY, J. L., AND W. C. REEVES. 1990a. Experimental studies on infection in vectors. Pages 145–253 *in* Epidemiology and Control of Mosquito-borne Arboviruses in California, 1943–1987 (W. C. Reeves, Ed.). California Mosquito and Vector Control Association, Sacramento.

HARDY, J. L., AND W. C. REEVES. 1990b. Experimental studies on infection in vertebrate hosts. Pages 66–127 *in* Epidemiology and Control of Mosquito-borne Arboviruses in California, 1943–1987 (W. C. Reeves, Ed.). California Mosquito and Vector Control Association, Sacramento.

HARTEMINK, N. A., S. A. DAVIS, P. REITER, Z. HUBÁLEK, AND J. A. P. HEESTERBEEK. 2007. Importance of bird-to-bird transmission for the establishment of West Nile virus. Vector-Borne and Zoonotic Diseases 7:575–584.

HASSAN, H. K., E. W. CUPP, G. E. HILL, C. R. KATHOLI, K. KLINGLER, AND T. R. UNNASCH. 2003. Avian host preference by vectors of eastern equine encephalomyelitis virus. American Journal of Tropical Medicine and Hygiene 69:641–647.

HAYES, C. G. 1989. West Nile fever. Pages 59–88 *in* The Arboviruses: Epidemiology and Ecology, vol. 5 (T. P. Monath, Ed.). CRC Press, Boca Raton, Florida.

HAYES, R. O., J. B. DANIELS, K. S. ANDERSON, M. A. PARSONS, H. K. MAXFIELD, AND L. C. LAMOTTE. 1962. Detection of eastern equine encephalitis virus and antibody in wild and domestic birds in Massachusetts. American Journal of Hygiene 75:183–189.

HAYES, R. O., D. B. FRANCY, J. S. LAZUICK, G. C. SMITH, AND E. P. J. GIBBS. 1977. Role of the cliff swallow bug (*Oeciacus vicarius*) in the natural cycle of a western equine encephalitis-related alphavirus. Journal of Medical Entomology 14:257–262.

HENDERSON, L. P., AND R. A. BRUST. 1977. Studies on potential transovarial transmission of western equine encephalomyelitis virus by *Culex tarsalis* Coquillett. Manitoba Entomologist 11:69–73.

HILDÉN, O., AND P. SAUROLA. 1982. Speed of autumn migration of birds ringed in Finland. Ornis Fennica 59:140–143.

HILL, G. E. 1993. House Finch (*Carpodacus mexicanus*). *In* The Birds of North America, no. 46 (A. Poole and F. Gill, Eds.). Academy of Natural Sciences, Philadelphia, and American Ornithologists' Union, Washington, D.C.

HODGSON, J. C., A. SPIELMAN, N. KOMAR, C. F. KRAHFORST, G. T. WALLACE, AND R. J. POLLACK. 2001. Interrupted blood-feeding by *Culiseta melanura* (Diptera: Culicidae) on European Starlings. Journal of Medical Entomology 38:59–66.

HOLDEN, P. 1955a. Recovery of western equine encephalomyelitis virus from naturally infected English Sparrows of New Jersey, 1953. Proceedings of the Society of Experimental Biology and Medicine 88:490–492.

HOLDEN, P. 1955b. Transmission of eastern equine encephalomyelitis in Ring-necked Pheasants. Proceedings of the Society of Experimental Biology and Medicine 88:607–610.

HOLDEN, P., R. O. HAYES, C. J. MITCHELL, D. B. FRANCY, J. S. LAZUICK, AND T. B. HUGHES. 1973. House Sparrows, *Passer domesticus* (L.), as hosts of arboviruses in Hale County, Texas. I. Field studies, 1965–1969. American Journal of Tropical Medicine and Hygiene 22:244–253.

HOOGSTRAAL, H., M. N. KAISER, M. A. TRAYLOR, S. GABER, AND E. GUINDY. 1961. Ticks (Ixodoidea) on birds migrating from Africa to Europe and Asia. Bulletin of the World Health Organization 24:197–212.

HOPLA, C. E., D. B. FRANCY, C. H. CALISHER, AND J. S. LAZUICK. 1993. Relationship of Cliff Swallows, ectoparasites, and an alphavirus in west-central Oklahoma. Journal of Medical Entomology 30:267–272.

HOWARD, J. J., M. A. GRAYSON, D. J. WHITE, AND J. OLIVER. 1996. Evidence for multiple foci of eastern equine encephalitis virus (Togaviridae: *Alphavirus*) in central New York State. Journal of Medical Entomology 33:421–432.

HOWARD, J. J., J. OLIVER, AND M. A. GRAYSON. 2004. Antibody response of wild birds to natural infection with *Alphaviruses*. Journal of Medical Entomology 41:1090–1103.

HOWARD, J. J., D. J. WHITE, AND S. L. MULLER. 1989. Mark–recapture studies on the *Culiseta* (Diptera: Culicidae) vectors of eastern equine encephalitis virus. Journal of Medical Entomology 26:190–199.

HUBÁLEK, Z. 2004. An annotated checklist of pathogenic microorganisms associated with migratory birds. Journal of Wildlife Diseases 40:639–659.

HUBÁLEK, Z. 2008. Mosquito-borne viruses in Europe. Parasitology Research 103 (Supplement 1):S29–S43.

HULL, J., A. HULL, W. REISEN, Y. FANG, AND H. ERNEST. 2006. Variation of West Nile virus antibody prevalence in migrating and wintering hawks in central California. Condor 108:435–439.

HÜPPOP, O., AND W. WINKEL. 2006. Climate change and timing of spring migration in the long-distance migrant *Ficedula hypoleuca* in central Europe: The role of spatially different temperature changes along migration routes. Journal of Ornithology 147:344–353.

HUYVAERT, K. P., A. T. MOORE, N. A. PANELLA, E. A. EDWARDS, M. B. BROWN, N. KOMAR, AND C. R. BROWN. 2008. Experimental inoculation of House Sparrows (*Passer domesticus*) with Buggy Creek virus. Journal of Wildlife Diseases 44:331–340.

JOHANSEN, C. A., D. J. NISBET, P. N. FOLEY, A. F. VAN DEN HURK, R. A. HALL, J. S. MACKENZIE, AND S. A. RITCHIE. 2004. Flavivirus isolations from mosquitoes collected from Saibai Island in the Torres Strait, Australia, during an incursion of Japanese encephalitis virus. Medical and Veterinary Entomology 18:281–287.

JOHNSON, C. G. 1969. Migration and Dispersal of Insects by Flight. Methuen, London.

JOHNSON, H. N. 1960. Public health in relation to birds: Arthropod-borne viruses. Transactions of the North American Wildlife and Natural Resources Conference 25:121–133.

JOHNSON, H. N., AND J. CASALS. 1972. Arboviruses associated with marine bird colonies of the Pacific Ocean region. Pages 200–204 *in* Transcontinental Connections of Migratory Birds and Their Role in the Distribution of Arboviruses. Nauka, Academy of Sciences of the USSR, Siberian Branch, Biological Institute, Novosibirsk, Siberia (USSR).

JOHNSON, O. W., C. D. ADLER, L. A. AYRES, M. A. BISHAP, J. E. DOSTER, P. M. JOHNSON, R. J. KIENHOLZ, AND S. E. SAVAGE. 2004. Radio-tagged Pacific Golden-Plovers: Further insight concerning the Hawaii–Alaska migratory link. Wilson Bulletin 116:158–162.

JOURDAIN, E., M. GAUTHIER-CLERC, D. J. BICOUT, AND P. SABATIER. 2007a. Bird migration routes and risk for pathogen dispersion into western Mediterranean wetlands. Emerging Infectious Diseases 13:365–372.

JOURDAIN, E., Y. TOUSSAINT, A. LEBLOND, D. J. BICOUT, P. SABATIER, AND M. GAUTHIER-CLERC. 2007b. Bird species potentially involved in introduction, amplification, and spread of West Nile virus in a Mediterranean wetland, the Camargue (southern France). Vector-Borne and Zoonotic Diseases 7:15–33.

JOURDAIN, E., H. G. ZELLER, P. SABATIER, S. MURRI, Y. KAYSER, T. GREENLAND, M. LAFAYE, AND M. GAUTHIER-CLERC. 2008. Prevalence of West Nile virus neutralizing antibodies in wild birds from the Camargue area, southern France. Journal of Wildlife Diseases 44:766–771.

Julian, K. G., M. Eidson, A. M. Kipp, E. Weiss, L. R. Petersen, J. R. Miller, S. R. Hinten, and A. A. Marfin. 2002. Early season crow mortality as a sentinel for West Nile virus disease in humans, northeastern United States. Vector-Borne and Zoonotic Diseases 2:145–155.

Kafuko, G. W. 1972. A review of studies of birds in relation to arboviruses known to occur in East Africa. Pages 173–176 in Transcontinental Connections of Migratory Birds and their Role in the Distribution of Arboviruses. Nauka, Academy of Sciences of the USSR, Siberian Branch, Biological Institute, Novosibirsk, Siberia (USSR).

Kay, B. H., and R. A. Farrow. 2000. Mosquito (Diptera: Culicidae) dispersal: Implications for the epidemiology of Japanese and Murray Valley encephalitis viruses in Australia. Journal of Medical Entomology 37:797–801.

Kilpatrick, A. M., A. A. Chmura, D. W. Gibbons, R. C. Fleischer, P. P. Marra, and P. Daszak. 2006a. Predicting the global spread of H5N1 avian influenza. Proceedings of the National Academy of Sciences USA 103:19368–19373.

Kilpatrick, A. M., P. D. Daszak, M. J. Jones, P. P. Marra, and L. D. Kramer. 2006b. Host heterogeneity dominates West Nile virus transmission. Proceedings of the Royal Society of London, Series B 273:2327–2333.

Kilpatrick, A. M., Y. Gluzberg, J. Burgett, and P. Daszak. 2004. Quantitative risk assessment of the pathways by which West Nile virus could reach Hawaii. EcoHealth 1:205–209.

Kilpatrick, A. M., L. D. Kramer, M. J. Jones, P. P. Marra, and P. Daszak. 2006c. West Nile virus epidemics in North America are driven by shifts in mosquito feeding behavior. PLoS Biology 4:606–610.

Kilpatrick, A. M., S. L. LaDeau, and P. P. Marra. 2007. Ecology of West Nile virus transmission and its impact on birds in the Western Hemisphere. Auk 124:1121–1136.

Kissling, R. E., R. W. Chamberlain, D. B. Nelson, and D. D. Stamm. 1955. Studies on the North American arthropod-borne encephalitides. VIII. Equine encephalitis studies in Louisiana. American Journal of Hygiene 62:233–254.

Kissling, R. E., R. W. Chamberlain, R. K. Sikes, and M. E. Eidson. 1954. Studies on the North American arthropod-borne encephalitides. III. Eastern equine encephalitis in wild birds. American Journal of Hygiene 60:251–265.

Kissling, R. E., R. W. Chamberlain, W. D. Sudia, and D. D. Stamm. 1957. Western equine encephalitis in wild birds. American Journal of Hygiene 66:48–55.

Kissling, R. E., H. Rubin, R. W. Chamberlain, and M. E. Eidson. 1951. Recovery of virus of eastern equine encephalomyelitis from blood of a Purple Grackle. Proceedings of the Society of Experimental Biology and Medicine 77:398–399.

Kleijn, D., V. J. Munster, B. S. Ebbinge, D. A. Jonkers, G. J. D. M. Müskens, Y. Van Randen, and R. A. M. Fouchier. 2010. Dynamics and ecological consequences of avian influenza virus infection in Greater White-fronted Geese in their winter staging areas. Proceedings of the Royal Society of London, Series B 277:2041–2048.

Koenig, W. D., and J. T. Stahl. 2007. Late summer and fall nesting in the Acorn Woodpecker and other North American terrestrial birds. Condor 109:334–350.

Kokernot, R. H., J. Hayes, R. L. Will, C. H. Tempelis, D. H. M. Chan, and B. Radivojevic. 1969. Arbovirus studies in the Ohio-Mississippi Basin, 1964–1967. II. St. Louis encephalitis virus. American Journal of Tropical Medicine and Hygiene 18:750–761.

Kokernot, R. H., and B. M. McIntosh. 1959. Isolation of West Nile virus from a naturally infected human being and from a bird, Sylvietta rufescens (Vieillot). South African Medical Journal 33:987–989.

Kokko, H. 1999. Competition for early arrival in migratory birds. Journal of Animal Ecology 68:940–950.

Komar, N. 1997. Reservoir Capacity of Communally Roosting Passerine Birds for Eastern Equine Encephalitis (EEE) Virus. Ph.D. dissertation, Harvard University School of Public Health, Boston, Massachusetts.

Komar, N. 2003. West Nile virus: Epidemiology and ecology in North America. Advances in Virus Research 61:185–234.

Komar, N., and G. G. Clark. 2006. West Nile virus activity in Latin America and the Caribbean. Pan American Journal of Public Health 19:112–117.

Komar, N., D. J. Dohm, M. J. Turell, and A. Spielman. 1999. Eastern equine encephalitis virus in birds: Relative competence of European Starlings (Sturnus vulgaris). American Journal of Tropical Medicine and Hygiene 60:387–391.

Komar, N., R. Lanciotti, R. Bowen, S. Langevin, and M. Bunning. 2002. Detection of West Nile virus in oral and cloacal swabs collected from bird carcasses. Emerging Infectious Diseases 8:741–742.

Komar, N., S. Langevin, S. Hinten, N. Nemeth, E. Edwards, D. Hettler, B. Davis, R. Bowen, and M. Bunning. 2003. Experimental infection of North American birds with the New York 1999 strain of West Nile virus. Emerging Infectious Diseases 9:311–322.

Komar, N., N. A. Panella, S. A. Langevin, A. C. Brault, M. Amador, E. Edwards, and J. C. Owen. 2005. Avian hosts for West Nile virus in St. Tammany Parish, Louisiana, 2002. American Journal of Tropical Medicine and Hygiene 73:1031–1037.

Komar, O., M. B. Robbins, K. Klenk, B. J. Blitvich, N. L. Marlenee, K. L. Burkhalter, D. J. Gubler, G. Gonzálvez, C. J. Peña, A. T. Peterson, and N. Komar. 2003. West Nile virus transmission in resident birds, Dominican Republic. Emerging Infectious Diseases 9:1299–1302.

KORENBERG, E. I. 2000. Seasonal population dynamics of *Ixodes* ticks and tick-borne encephalitis virus. Experimental and Applied Acarology 24:665–681.

KRAMER, L. D., AND H. M. FALLAH. 1999. Genetic variation among isolates of western equine encephalomyelitis virus from California. American Journal of Tropical Medicine and Hygiene 60:708–713.

KRAMER, L. D., S. B. PRESSER, J. L. HARDY, AND A. O. JACKSON. 1997. Genotypic and phenotypic variation of selected Saint Louis encephalitis viral strains isolated in California. American Journal of Tropical Medicine and Hygiene 57:222–229.

KRAMER, L. D., L. M. STYER, AND G. D. EBEL. 2008. A global perspective on the epidemiology of West Nile virus. Annual Review of Entomology 53:61–81.

KRAMER, L. D., T. M. WOLFE, E. N. GREEN, R. E. CHILES, H. FALLAH, Y. FANG, AND W. K. REISEN. 2002. Detection of encephalitis viruses in mosquitoes (Diptera: Culicidae) and avian tissues. Journal of Medical Entomology 39:312–323.

KRAUSS, S., C. A. OBERT, J. FRANKS, D. WALKER, K. JONES, P. SEILER, L. NILES, S. P. PRYOR, J. C. OBENAUER, C. W. NAEVE, AND OTHERS. 2007. Influenza in migratory birds and evidence of limited intercontinental virus exchange. PLoS Pathogens 3:1684–1693.

KUNO, G. 2001. Persistence of arboviruses and antiviral antibodies in vertebrate hosts: Its occurrence and impacts. Reviews in Medical Virology 11:165–190.

KURKELA, S., O. RÄTTI, E. HUHTAMO, N. Y. UZCÁTEGUI, J. P. NUORTI, J. LAAKKONEN, T. MANNI, P. HELLE, A. VAHERI, AND O. VAPALAHTI. 2008. Sindbis virus infection in resident birds, migratory birds, and humans, Finland. Emerging Infectious Diseases 14:41–47.

KWAN, J. L., S. KLUH, M. B. MADON, AND W. K. REISEN. 2010. West Nile virus emergence and persistence in Los Angeles, California, 2003–2008. American Journal of Tropical Medicine and Hygiene 83:400–412.

LACK, D. 1968. Ecological Adaptations for Breeding in Birds. Methuen, London.

LADEAU, S. L., A. M. KILPATRICK, AND P. P. MARRA. 2007. West Nile virus emergence and large-scale declines of North American bird populations. Nature 447:710–713.

LAINE, M., R. LUUKKAINEN, AND A. TOIVANEN. 2004. Sindbis viruses and other alphaviruses as cause of human disease. Journal of Internal Medicine 256:457–471.

LAMBERT, A. J., D. A. MARTIN, AND R. S. LANCIOTTI. 2003. Detection of North American eastern and western equine encephalitis viruses by nucleic acid amplification assays. Journal of Clinical Microbiology 41:379–385.

LANCIOTTI, R. S., A. J. KERST, R. S. NASCI, M. S. GODSEY, C. J. MITCHELL, H. M. SAVAGE, N. KOMAR, N. A. PANELLA, B. C. ALLEN, K. E. VOLPE, AND OTHERS. 2000. Rapid detection of West Nile virus from human clinical specimens, field-collected mosquitoes, and avian samples by a TaqMan reverse transcriptase-PCR assay. Journal of Clinical Microbiology 38:4066–4071.

LANGEVIN, S. A., A. C. BRAULT, N. A. PANELLA, R. A. BOWEN, AND N. KOMAR. 2005. Variation in virulence of West Nile virus strains for House Sparrows (*Passer domesticus*). American Journal of Tropical Medicine and Hygiene 72:99–102.

LASALLE, M. W., AND M. E. DAKIN. 1982. Dispersal of *Culex salinarius* in southwestern Louisiana. Mosquito News 42:543–550.

LEBARBENCHON, C., C. J. FEARE, F. RENAUD, F. THOMAS, AND M. GAUTHIER-CLERC. 2010. Persistence of highly pathogenic avian influenza viruses in natural ecosystems. Emerging Infectious Diseases 16:1057–1062.

LEE, J.-K., W. A. ROWLEY, AND K. B. PLATT. 2000. Longevity and spontaneous flight activity of *Culex tarsalis* (Diptera: Culicidae) infected with western equine encephalomyelitis virus. Journal of Medical Entomology 37:187–193.

LETCHWORTH, G. J., J. DEL C. BARRERA, J. R. FISHEL, AND L. RODRIGUEZ. 1996. Vesicular stomatitis New Jersey virus RNA persists in cattle following convalescence. Virology 219:480–484.

LEVINE, B., J. M. HARDWICK, AND D. E. GRIFFIN. 1994. Persistence of alphaviruses in vertebrate hosts. Trends in Microbiology 2:25–28.

LEWIS, M., J. RENCŁAWOWICZ, AND P. VAN DEN DRIESSCHE. 2006. Traveling waves and spread rates for a West Nile virus model. Bulletin of Mathematical Biology 68:3–23.

LILLIE, T. H., W. C. MARQUARDT, AND R. H. JONES. 1981. The flight range of *Culicoides variipennis* (Diptera: Ceratopogonidae). Canadian Entomologist 113:419–426.

LINDSTRÖM, K., I. T. VAN DER VEEN, B.-A. LEGAULT, AND J. O. LUNDSTRÖM. 2003. Activity and predator escape performance of Common Greenfinches *Carduelis chloris* infected with Sindbis virus. Ardea 91:103–111.

LINKE, S., M. NIEDRIG, A. KAISER, H. ELLERBROK, K. MÜLLER, T. MÜLLER, F. J. CONRATHS, R.-U. MÜHLE, D. SCHMIDT, AND OTHERS. 2007. Serological evidence of West Nile virus infections in wild birds captured in Germany. American Journal of Tropical Medicine and Hygiene 77:358–364.

LÓPEZ, G., M. A. JIMÉNEZ-CLAVERO, C. G. TEJEDOR, R. SORIGUER, AND J. FIGÚEROLA. 2008. Prevalence of West Nile virus neutralizing antibodies in Spain is related to the behavior of migratory birds. Vector-Borne and Zoonotic Diseases 8:615–621.

LORD, R. D., AND C. H. CALISHER. 1970. Further evidence of southward transport of arboviruses by migratory birds. American Journal of Epidemiology 92:73–78.

LORD, R. D., C. H. CALISHER, W. A. CHAPPELL, W. R. METZGER, AND G. W. FISCHER. 1974. Urban St. Louis encephalitis surveillance through wild birds. American Journal of Epidemiology 99:360–363.

Loss, S. R., G. L. Hamer, T. L. Goldberg, M. O. Ruiz, U. D. Kitron, E. D. Walker, and J. D. Brawn. 2009. Nestling passerines are not important hosts for amplification of West Nile virus in Chicago, Illinois. Vector-Borne and Zoonotic Diseases 9:13–17.

Lowther, P. E., and C. L. Cink. 1992. House Sparrow (*Passer domesticus*). *In* The Birds of North America, no. 12 (A. Poole, P. Stettenheim, and F. Gill, Eds.). Academy of Natural Sciences, Philadelphia, and American Ornithologists' Union, Washington, D.C.

Ludwig, G. V., R. S. Cook, R. G. McLean, and D. B. Francy. 1986. Viremic enhancement due to transo-varially acquired antibodies to St. Louis encephalitis virus in birds. Journal of Wildlife Diseases 22:326–334.

Lundström, J. O. 1994. Vector competence of western European mosquitoes for arboviruses: A review of field and experimental studies. Bulletin of the Society of Vector Ecology 19:23–36.

Lundström, J. O. 1999. Mosquito-borne viruses in western Europe: A review. Journal of Vector Ecology 24:1–39.

Lundström, J. O., K. M. Lindström, B. Olsen, D. Krakower, and R. Dufa. 2001. Prevalence of Sindbis virus neutralizing antibodies among Swedish passerines indicates that thrushes are the main amplifying hosts. Journal of Medical Entomology 38:289–297.

Lundström, J. O., M. J. Turell, and B. Niklasson. 1993. Viremia in three orders of birds (Anseriformes, Galliformes and Passeriformes) inoculated with Ockelbo virus. Journal of Wildlife Diseases 29:189–195.

Lvov, D. K., A. M. Butenko, V. L. Gromashevsky, A. I. Kovtunov, A. G. Prilipov, R. Kinney, V. A. Aristova, A. F. Dzharkenov, E. I. Samokhvalov, H. M. Savage, and others. 2004. West Nile virus and other zoonotic viruses in Russia: Examples of emerging-reemerging situations. Pages 85–96 *in* Emergence and Control of Zoonotic Viral Encephalitides (C. H. Calisher and D. E. Griffin, Eds.). Springer, New York.

Maciel-de-Freitas, R., R. B. Neto, J. M. Concalves, C. T. Codeco, and R. Lourenco-de-Oliveira. 2006. Movement of dengue vectors between the human modified environment and an urban forest in Rio de Janeiro. Journal of Medical Entomology 43:1112–1120.

Mack, D. E., and W. Yong. 2000. Swainson's Thrush (*Catharus ustulatus*). *In* The Birds of North America, no. 540 (A. Poole and F. Gill, Eds.). Birds of North America, Philadelphia.

Madder, D. J., G. A. Surgeoner, and B. V. Helson. 1983. Number of generations, egg production, and developmental time of *Culex pipiens* and *Culex restuans* (Diptera: Culicidae) in southern Ontario. Journal of Medical Entomology 20:275–287.

Mahmood, F., R. E. Chiles, Y. Fang, C. M. Barker, and W. K. Reisen. 2004a. Role of nestling Mourning Doves and House Finches as amplifying hosts of St. Louis encephalitis virus. Journal of Medical Entomology 41:965–972.

Mahmood, F., W. K. Reisen, R. E. Chiles, and Y. Fang. 2004b. Western equine encephalomyelitis virus infection affects the life table characteristics of *Culex tarsalis* (Diptera: Culicidae). Journal of Medical Entomology 41:982–986.

Malkinson, M., and C. Banet. 2002. The role of birds in the ecology of West Nile virus in Europe and Africa. Current Topics in Microbiology and Immunology 267:309–322.

Malkinson, M., C. Banet, Y. Weisman, S. Pokamunski, R. King, M.-T. Drouet, and V. Deubel. 2002. Introduction of West Nile virus in the Middle East by migrating White Storks. Emerging Infectious Diseases 8:392–397.

Marra, P. P., S. Griffing, C. Caffrey, A. M. Kilpatrick, R. McLean, C. Brand, E. Saito, A. P. Dupuis, L. Kramer, and R. Novak. 2004. West Nile virus and wildlife. BioScience 54:393–402.

McLean, R. G. 2006. West Nile virus in North American birds. Pages 44–64 *in* Current Topics in Avian Disease Research: Understanding Endemic and Invasive Diseases (R. K. Barraclough, Ed.). Ornithological Monographs, no. 60.

McLean, R. G., and G. S. Bowen. 1980. Vertebrate hosts. Pages 381–450 *in* St. Louis Encephalitis (T. P. Monath, Ed.). American Public Health Association, Washington, D.C.

McLean, R. G., W. J. Crans, D. F. Caccamise, J. McNelly, L. J. Kirk, C. J. Mitchell, and C. H. Calisher. 1995. Experimental infection of wading birds with eastern equine encephalitis virus. Journal of Wildlife Diseases 31:502–508.

McLean, R. G., G. Frier, G. L. Parham, D. B. Francy, T. P. Monath, E. G. Campos, A. Thierrien, J. Kerschner, and C. H. Calisher. 1985. Investigations of the vertebrate hosts of eastern equine encephalitis during an epizootic in Michigan, 1980. American Journal of Tropical Medicine and Hygiene 34:1190–1202.

McLean, R. G., L. J. Kirk, R. B. Shriner, and M. Townsend. 1993. Avian hosts of St. Louis encephalitis virus in Pine Bluff, Arkansas, 1991. American Journal of Tropical Medicine and Hygiene 49:46–52.

McLean, R. G., J. Mullenix, J. Kerschner, and J. Hamm. 1983. The House Sparrow (*Passer domesticus*) as a sentinel host for St. Louis encephalitis virus. American Journal of Tropical Medicine and Hygiene 32:1120–1129.

McLean, R. G., and T. W. Scott. 1979. Avian hosts of St. Louis encephalitis virus. Proceedings of the Bird Control Seminars 8:143–155.

McLean, R. G., S. R. Ubico, D. E. Docherty, W. R. Hansen, L. Sileo, and T. S. McNamara. 2001. West Nile virus transmission and ecology in birds. Annals of the New York Academy of Sciences 951:54–57.

MELVILLE, D. S., AND K. F. SHORTRIDGE. 2006. Spread of H5N1 avian influenza virus: An ecological conundrum. Letters in Applied Microbiology 42:435–437.

MERRILL, S. A., F. B. RAMBERG, AND H. H. HAGEDORN. 2005. Phylogeography and population structure of *Aedes aegypti* in Arizona. American Journal of Tropical Medicine and Hygiene 72:304–310.

MILBY, M. M., AND W. C. REEVES. 1990. Natural infection in vertebrate hosts other than man. Pages 26–66 *in* Epidemiology and Control of Mosquito-borne Arboviruses in California, 1943–1987 (W. C. Reeves, Ed.). California Mosquito and Vector Control Association, Sacramento.

MILBY, M. M., W. K. REISEN, AND W. C. REEVES. 1983. Intercanyon movement of marked *Culex tarsalis* (Diptera: Culicidae). Journal of Medical Entomology 20:193–198.

MILES, J. A. R., AND D. W. HOWES. 1953. Observations on virus encephalitis in South Australia. Medical Journal of Australia 1:7–12.

MIMS, C. A. 1981. Vertical transmission of viruses. Microbiology Reviews 45:267–286.

MIRARCHI, R. E., AND T. S. BASKETT. 1994. Mourning Dove (*Zenaida macroura*). *In* The Birds of North America, no. 117 (A. Poole and F. Gill, Eds.). Academy of Natural Sciences, Philadelphia, and American Ornithologists' Union, Washington, D.C.

MOLAEI, G., T. G. ANDREADIS, P. M. ARMSTRONG, AND M. DIUK-WASSER. 2008. Host-feeding patterns of potential mosquito vectors in Connecticut, USA: Molecular analysis of bloodmeals from 23 species of *Aedes, Anopheles, Culex, Coquillettidia, Psorophora,* and *Uranotaenia.* Journal of Medical Entomology 45:1143–1151.

MØLLER, A. P., AND J. ERRITZØE. 1998. Host immune defence and migration in birds. Evolutionary Ecology 12:945–953.

MONATH, T. P., ED. 1980. St. Louis Encephalitis. American Public Health Association, Washington, D.C.

MONCAYO, A. C., J. D. EDMAN, AND M. J. TURELL. 2000. Effect of eastern equine encephalomyelitis virus on the survival of *Aedes albopictus, Anopheles quadrimaculatus,* and *Coquillettidia perturbans* (Diptera: Culicidae). Journal of Medical Entomology 37:701–706.

MOORE, A. T., E. A. EDWARDS, M. B. BROWN, N. KOMAR, AND C. R. BROWN. 2007. Ecological correlates of Buggy Creek virus infection in *Oeciacus vicarius,* southwestern Nebraska, 2004. Journal of Medical Entomology 44:42–49.

MOORE, F., AND P. KERLINGER. 1987. Stopover and fat deposition by North American wood-warblers (Parulinae) following spring migration over the Gulf of Mexico. Oecologia 74:47–54.

MORRIS, C. D. 1988. Eastern equine encephalomyelitis. Pages 1–20 *in* The Arboviruses: Epidemiology and Ecology, vol. 3 (T. P. Monath, Ed.). CRC Press, Boca Raton, Florida.

MORRIS, C. D., M. E. COREY, D. E. EMORD, AND J. J. HOWARD. 1980. Epizootiology of eastern equine encephalomyelitis virus in upstate New York, USA. I. Introduction, demography and natural environment of an endemic focus. Journal of Medical Entomology 17:442–452.

MORRIS, C. D., AND S. SRIHONGSE. 1978. An evaluation of the hypothesis of transovarial transmission of eastern equine encephalomyelitis virus by *Culiseta melanura.* American Journal of Tropical Medicine and Hygiene 27:1246–1250.

MORRIS, C. D., E. WHITNEY, T. F. BAST, AND R. DEIBEL. 1973. An outbreak of eastern equine encephalomyelitis in upstate New York during 1971. American Journal of Tropical Medicine and Hygiene 22:561–566.

MORSHED, M. G., J. D. SCOTT, K. FERNANDO, L. BEATI, D. F. MAZEROLLE, G. GEDDES, AND L. A. DURDEN. 2005. Migratory songbirds disperse ticks across Canada, and first isolation of the Lyme disease spirochete, *Borrelia burgdorferi,* from the avian tick, *Ixodes auritulus.* Journal of Parasitology 91:780–790.

NASCI, R. S., N. KOMAR, A. A. MARFIN, G. V. LUDWIG, L. D. KRAMER, T. J. DANIELS, R. C. FALCO, S. R. CAMPBELL, K. BROOKES, K. L. GOTTFRIED, AND OTHERS. 2002. Detection of West Nile virus-infected mosquitoes and seropositive juvenile birds in the vicinity of virus-positive dead birds. American Journal of Tropical Medicine and Hygiene 67:492–496.

NEMETH, N. M., S. BECKETT, E. EDWARDS, K. KLENK, AND N. KOMAR. 2007. Avian mortality surveillance for West Nile virus in Colorado. American Journal of Tropical Medicine and Hygiene 76:431–437.

NEMETH, N., D. GOULD, R. BOWEN, AND N. KOMAR. 2006a. Natural and experimental West Nile virus infection in five raptor species. Journal of Wildlife Diseases 42:1–13.

NEMETH, N. M., D. C. HAHN, D. H. GOULD, AND R. A. BOWEN. 2006b. Experimental West Nile virus infection in Eastern Screech Owls (*Megascops asio*). Avian Diseases 50:252–258.

NEMETH, N., G. YOUNG, C. NDALUKA, H. BIELEFELDT-OHMANN, N. KOMAR, AND R. BOWEN. 2009. Persistent West Nile virus infection in the House Sparrow (*Passer domesticus*). Archives of Virology 154:783–789.

NGA, P. T., M. DEL C. PARQUET, V. D. CUONG, S.-P. MA, F. HASEBE, S. INOUE, Y. MAKINO, M. TAKAGI, V. S. NAM, AND K. MORITA. 2004. Shift in Japanese encephalitis virus (JEV) genotype circulating in northern Vietnam: Implications for frequent introductions of JEV from southeast Asia to east Asia. Journal of General Virology 85:1625–1631.

NGO, K. A., AND L. D. KRAMER. 2003. Identification of mosquito bloodmeals using polymerase chain reaction (PCR) with order-specific primers. Journal of Medical Entomology 40:215–222.

NICE, C. S. 1994. The dissemination of human infectious disease by birds. Reviews in Medical Microbiology 5:191–198.

NIELSEN, C. F., AND W. K. REISEN. 2007. West Nile-infected dead corvids increase the risk of infection in *Culex* mosquitoes (Diptera: Culicidae) in domestic landscapes. Journal of Medical Entomology 44:1067–1073.

NIKLASSON, B. 1989. Sindbis and Sindbis-like viruses. Pages 167–176 *in* The Arboviruses: Epidemiology and Ecology, vol. 4 (T. P. Monath, Ed.). CRC Press, Boca Raton, Florida.

NIR, Y., R. GOLDWASSER, Y. LASOWSKI, AND A. AVIVI. 1967. Isolation of arboviruses from wild birds in Israel. American Journal of Epidemiology 86:372–378.

NORDER, H., J. O. LUNDSTRÖM, O. KOZUCH, AND L. O. MAGNIUS. 1996. Genetic relatedness of Sindbis virus strains from Europe, Middle East, and Africa. Virology 222:440–445.

OBERHOLSER, H. C. 1974. The Bird Life of Texas, vol. 1. University of Texas Press, Austin.

O'BRIEN, V. A., C. U. METEYER, H. S. IP, R. R. LONG, AND C. R. BROWN. 2010a. Pathology and virus detection in tissues of nestling House Sparrows naturally infected with Buggy Creek virus (Togaviridae). Journal of Wildlife Diseases 46:23–32.

O'BRIEN, V. A., C. U. METEYER, W. K. REISEN, H. S. IP, AND C. R. BROWN. 2010b. Prevalence and pathology of West Nile virus in naturally infected House Sparrows, western Nebraska, 2008. American Journal of Tropical Medicine and Hygiene 82:937–944.

O'BRIEN, V. A., A. T. MOORE, K. P. HUYVAERT, AND C. R. BROWN. 2008. No evidence for spring reintroduction of an arbovirus by Cliff Swallows. Wilson Journal of Ornithology 120:910–913.

O'BRIEN, V. A., A. T. MOORE, G. R. YOUNG, N. KOMAR, W. K. REISEN, AND C. R. BROWN. 2011. An enzootic vector-borne virus is amplified at epizootic levels by an invasive avian host. Proceedings of the Royal Society of London, Series B 278:239–246.

OKUBO, A. 1980. Diffusion and Ecological Problems: Mathematical Models. Springer-Verlag, Berlin.

OLSEN, B., V. J. MUNSTER, A. WALLENSTEN, J. WALDENSTRÖM, A. D. M. OSTERHAUS, AND R. A. M. FOUCHIER. 2006. Global patterns of influenza A virus in wild birds. Science 312:384–388.

OWEN, J. [C.], F. MOORE, N. PANELLA, E. EDWARDS, R. BRU, M. HUGHES, AND N. KOMAR. 2006. Migrating birds as dispersal vehicles for West Nile virus. EcoHealth 3:79–85.

OWEN, J. C., F. R. MOORE, A. J. WILLIAMS, M. P. WARD, T. A. BEVEROTH, E. A. MILLER, L. C. WILSON, V. J. MORLEY, R. N. ABBEY-LEE, B. A. VEENEMAN, AND OTHERS. 2010. Test of recrudescence hypothesis for overwintering of West Nile virus in Gray Catbirds. Journal of Medical Entomology 47:451–457.

PADHI, A., A. T. MOORE, M. B. BROWN, J. E. FOSTER, M. PFEFFER, AND C. R. BROWN. 2011. Isolation by distance explains genetic structure of Buggy Creek virus, a bird-associated arbovirus. Evolutionary Ecology 25:403–416.

PADHI, A., A. T. MOORE, M. B. BROWN, J. E. FOSTER, M. PFEFFER, K. P. GAINES, V. A. O'BRIEN, S. A. STRICKLER, A. E. JOHNSON, AND C. R. BROWN. 2008. Phylogeographical structure and evolutionary history of two Buggy Creek virus lineages in the western Great Plains of North America. Journal of General Virology 89:2122–2131.

PARTNERS IN FLIGHT. 2007. PIF Landbird Population Estimates Database. Rocky Mountain Bird Observatory, Fort Collins, Colorado. [Online.] Available at www.rmbo.org/pif_db/laped/.

PATZ, J. A., AND W. K. REISEN. 2001. Immunology, climate change and vector-borne diseases. Trends in Immunology 22:171–172.

PAVRI, K. M., F. M. RODRIGUES, AND P. K. RAJAGOPALAN. 1972. Role of birds in the ecology of arboviruses encountered in India. Pages 193–200 *in* Transcontinental Connections of Migratory Birds and Their Role in the Distribution of Arboviruses. Nauka, Academy of Sciences of the USSR, Siberian Branch, Biological Institute, Novosibirsk, Siberia (USSR).

PEARCE, J. M., A. M. RAMEY, P. L. FLINT, A. V. KOEHLER, J. P. FLESKES, J. C. FRANSON, J. S. HALL, D. V. DERKSEN, AND H. S. IP. 2009. Avian influenza at both ends of a migratory flyway: Characterizing viral genomic diversity to optimize surveillance plans for North America. Evolutionary Applications 2:457–468.

PEDGLEY, D. E. 1983. Windborne spread of insect-transmitted diseases of animals and man. Philosophical Transactions of the Royal Society of London, Series B 302:463–470.

PETERSEN, L. R., AND J. T. ROEHRIG. 2001. West Nile virus: A reemerging global pathogen. Emerging Infectious Diseases 7:611–614.

PETERSON, A. T., D. A. VIEGLAIS, AND J. K. ANDREASEN. 2003. Migratory birds modeled as critical transport agents for West Nile virus in North America. Vector-Borne and Zoonotic Diseases 3:27–37.

PFEFFER, M., J. E. FOSTER, E. A. EDWARDS, M. B. BROWN, N. KOMAR, AND C. R. BROWN. 2006. Phylogenetic analysis of Buggy Creek virus: Evidence for multiple clades in the western Great Plains, United States of America. Applied and Environmental Microbiology 72:6886–6893.

PIERRE, J. P., AND H. HIGUCHI. 2004. Satellite tracking in avian conservation: Applications and results from Asia. Memoirs of the National Institute of Polar Research, Special Issue 58:101–109.

POWERS, A. M., A. C. BRAULT, Y. SHIRAKO, E. G. STRAUSS, W. KANG, J. H. STRAUSS, AND S. C. WEAVER. 2001. Evolutionary relationships and systematics of the alphaviruses. Journal of Virology 75:10118–10131.

PULIDO, F., AND P. BERTHOLD. 2010. Current selection for lower migratory activity will drive the evolution of residency in a migratory bird population. Proceedings of the National Academy of Sciences USA 107:7341–7346.

RAESS, M. 2008. Continental efforts: Migration speed in spring and autumn in an inner-Asian migrant. Journal of Avian Biology 39:13–18.

RAND, P. W., E. H. LACOMBE, R. P. SMITH, JR., AND J. FICKER. 1998. Participation of birds (Aves) in the emergence of Lyme disease in southern Maine. Journal of Medical Entomology 35:270–276.

RAPPOLE, J. H., B. W. COMPTON, P. LEIMBRUBER, J. ROBERTSON, D. I. KING, AND S. C. RENNER. 2006. Modeling movement of West Nile virus in the Western Hemisphere. Vector-Borne and Zoonotic Diseases 6:128–139.

RAPPOLE, J. H., S. R. DERRICKSON, AND Z. HUBÁLEK. 2000. Migratory birds and spread of West Nile virus in the Western Hemisphere. Emerging Infectious Diseases 6:319–328.

RAPPOLE, J. H., AND Z. HUBÁLEK. 2003. Migratory birds and West Nile virus. Journal of Applied Microbiology 94:47S–58S.

RAPPOLE, J. H., AND D. W. WARNER. 1976. Relationships between behavior, physiology, and weather in avian transients at a migration stopover site. Oecologia 26:193–212.

REED, K. D., J. K. MEECE, J. S. HENKEL, AND S. K. SHUKLA. 2003. Birds, migration and emerging zoonoses: West Nile virus, Lyme disease, influenza A and entero-pathogens. Clinical Medical Research 1:5–12.

REEVES, W. C. 1974. Overwintering of arboviruses. Progress in Medical Virology 17:193–220.

REEVES, W. C. 1990. Overwintering of arboviruses. Pages 357–382 *in* Epidemiology and Control of Mosquito-borne Arboviruses in California, 1943–1987 (W. C. Reeves, Ed.). California Mosquito and Vector Control Association, Sacramento.

REEVES, W. C., G. A. HUTSON, R. E. BELLAMY, AND R. P. SCRIVANI. 1958. Chronic latent infections of birds with western equine encephalomyelitis virus. Proceedings of the Society of Experimental Biology and Medicine 97:733–736.

REISEN, W. K. 1990. North American mosquito-borne arboviruses: Questions of persistence and amplification. Bulletin of the Society of Vector Ecology 15:11–21.

REISEN, W. K. 2003. Epidemiology of St. Louis encephalitis virus. Pages 139–183 *in* The Flaviviruses: Detection, Diagnosis and Vaccine Development (T. J. Chambers and T. P. Monath, Eds.). Elsevier Academic Press, San Diego, California.

REISEN, W. K., C. M. BARKER, R. CARNEY, H. D. LOTHROP, S. S. WHEELER, J. L. WILSON, M. B. MADON, R. TAKAHASHI, B. CARROLL, S. GARCIA, AND OTHERS. 2006a. Role of corvids in epidemiology of West Nile virus in southern California. Journal of Medical Entomology 43:356–367.

REISEN, W. K., J. P. BURNS, AND R. G. BASIO. 1972. A mosquito survey of Guam, Marianas Islands with notes on the vector borne disease potential. Journal of Medical Entomology 9:319–324.

REISEN, W. K., B. D. CARROLL, R. TAKAHASHI, Y. FANG, S. GARCIA, V. MARTINEZ, AND R. QUIRING. 2009a. Repeated West Nile virus epidemic transmission in Kern County, California, 2004–2007. Journal of Medical Entomology 46:139–157.

REISEN, W. K., R. E. CHILES, E. N. GREEN, Y. FANG, AND F. MAHMOOD. 2003a. Previous infection protects House Finches from re-infection with St. Louis encephalitis virus. Journal of Medical Entomology 40:300–305.

REISEN, W. K., R. E. CHILES, E. N. GREEN, Y. FANG, F. MAHMOOD, V. M. MARTINEZ, AND T. LAVER. 2003b. Effects of immunosuppression on encephalitis virus infection in the House Finch, *Carpodacus mexicanus*. Journal of Medical Entomology 40:206–214.

REISEN, W. K., R. E. CHILES, V. M. MARTINEZ, Y. FANG, AND E. N. GREEN. 2003c. Experimental infection of California birds with western equine encephalomyelitis and St. Louis encephalitis viruses. Journal of Medical Entomology 40:968–982.

REISEN, W. K., R. E. CHILES, V. M. MARTINEZ, Y. FANG, AND E. N. GREEN. 2004a. Encephalitis virus persistence in California birds: Experimental infections in Mourning Doves (*Zenaidura macroura*). Journal of Medical Entomology 41:462–466.

REISEN, W. K., Y. FANG, AND A. C. BRAULT. 2008. Limited interdecadal variation in mosquito (Diptera: Culicidae) and avian host competence for western equine encephalomyelitis virus (Togaviridae: *Alphavirus*). American Journal of Tropical Medicine and Hygiene 78:681–686.

REISEN, W. K., Y. FANG, H. D. LOTHROP, V. M. MARTINEZ, J. WILSON, P. O'CONNOR, R. CARNEY, B. CAHOON-YOUNG, M. SHAFII, AND A. C. BRAULT. 2006b. Overwintering of West Nile virus in southern California. Journal of Medical Entomology 43:344–355.

REISEN, W. K., Y. FANG, AND V. M. MARTINEZ. 2005a. Avian host and mosquito (Diptera: Culicidae) vector competence determine the efficiency of West Nile and St. Louis encephalitis transmission. Journal of Medical Entomology 42:367–375.

REISEN, W. K., J. L. HARDY, R. E. CHILES, L. D. KRAMER, V. M. MARTINEZ, AND S. B. PRESSER. 1996. Ecology of mosquitoes and lack of arbovirus activity at Morro Bay, San Luis Obispo County, California. Journal of the American Mosquito Control Association 12:679–687.

REISEN, W. K., L. D. KRAMER, R. E. CHILES, E. N. GREEN, AND V. M. MARTINEZ. 2001. Encephalitis virus persistence in California birds: Preliminary studies with House Finches. Journal of Medical Entomology 38:393–399.

REISEN, W. K., L. D. KRAMER, R. E. CHILES, V. M. MARTINEZ, AND B. F. ELDRIDGE. 2000a. Response of House Finches to infection with sympatric and allopatric strains of western equine encephalomyelitis and St. Louis encephalitis viruses from California. Journal of Medical Entomology 37:259–264.

REISEN, W. K., L. D. KRAMER, R. E. CHILES, T. M. WOLFE, AND E.-G. N. GREEN. 2002a. Simulated overwintering of encephalitis viruses in diapausing female *Culex tarsalis* (Diptera: Culicidae). Journal of Medical Entomology 39:226–233.

REISEN, W. K., H. D. LOTHROP, R. E. CHILES, R. CUSACK, E.-G. N. GREEN, Y. FANG, AND M. KENSINGTON. 2002b. Persistence and amplification of St. Louis encephalitis virus in the Coachella Valley of California, 2000–2001. Journal of Medical Entomology 39:793–805.

REISEN, W. K., H. LOTHROP, R. CHILES, M. MADON, C. COSSEN, L. WOODS, S. HUSTED, V. KRAMER, AND J. EDMAN. 2004b. West Nile virus in California. Emerging Infectious Diseases 10:1369–1378.

REISEN, W. K., J. O. LUNDSTROM, T. W. SCOTT, B. F. ELDRIDGE, R. E. CHILES, R. CUSACK, V. M. MARTINEZ, H. D. LOTHROP, D. GUTIERREZ, S. E. WRIGHT, AND OTHERS. 2000b. Patterns of avian seroprevalence to western equine encephalomyelitis and St. Louis encephalitis viruses in California, USA. Journal of Medical Entomology 37:507–527.

REISEN, W. K., AND T. P. MONATH. 1989. Western equine encephalomyelitis. Pages 89–137 *in* The Arboviruses: Epidemiology and Ecology, vol. 5 (T. P. Monath, Ed.). CRC Press, Boca Raton, Florida.

REISEN, W. K., AND W. C. REEVES. 1990. Bionomics and ecology of *Culex tarsalis* and other potential mosquito vector species. Pages 254–329 *in* Epidemiology and Control of Mosquito-borne Arboviruses in California, 1943–1987 (W. C. Reeves, Ed.). California Mosquito and Vector Control Association, Sacramento.

REISEN, W. K., S. WHEELER, M. V. ARMIJOS, Y. FANG, S. GARCIA, K. KELLEY, AND S. WRIGHT. 2009b. Role of communally nesting ardeid birds in the epidemiology of West Nile virus revisited. Vector-Borne and Zoonotic Diseases 9:275–280.

REISEN, W. K., S. S. WHEELER, S. GARCIA, AND Y. FANG. 2010. Migratory birds and the dispersal of arboviruses in California. American Journal of Tropical Medicine and Hygiene 83:808–815.

REISEN, W. K., S. S. WHEELER, S. YAMAMOTO, Y. FANG, AND S. GARCIA. 2005b. Nesting ardeid colonies are not a focus of elevated West Nile virus activity in southern California. Vector-Borne and Zoonotic Diseases 5:258–266.

RITCHIE, S. A., AND W. ROCHESTER. 2001. Wind-blown mosquitoes and introduction of Japanese encephalitis into Australia. Emerging Infectious Diseases 7:900–903.

ROSEN, L. 1981. Transovarial transmission of arboviruses by mosquitoes. Médecine Tropicale 41:23–29.

ROSEN, L. 1987. Overwintering mechanisms of mosquito-borne arboviruses in temperate climates. American Journal of Tropical Medicine and Hygiene 37 (Supplement):69S–76S.

RUSH, W. A., R. C. KENNEDY, AND C. M. EKLUND. 1963. Evidence against maintenance of western equine encephalomyelitis virus by *Culex tarsalis* during spring in northwestern United States. American Journal of Hygiene 77:258–264.

SALLABANKS, R., AND F. C. JAMES. 1999. American Robin (*Turdus migratorius*). *In* The Birds of North America, no. 462 (A. Poole and F. Gill, Eds.). Birds of North America, Philadelphia.

SAMMELS, L. M., M. D. LINDSAY, M. POIDINGER, R. J. COELEN, AND J. S. MACKENZIE. 1999. Geographic distribution and evolution of Sindbis virus in Australia. Journal of General Virology 80:739–748.

SCHERER, W. F., E. L. BUESCHER, AND H. E. MCCLURE. 1959. Ecologic studies of Japanese encephalitis virus in Japan. V. Avian factors. American Journal of Tropical Medicine and Hygiene 8:689–697.

SCHMALJOHANN, H., F. LIECHTI, AND B. BRUDERER. 2007a. An addendum to 'songbird migration across the Sahara: The non-stop hypothesis rejected!' Proceedings of the Royal Society of London, Series B 274:1919–1920.

SCHMALJOHANN, H., F. LIECHTI, AND B. BRUDERER. 2007b. Songbird migration across the Sahara: The non-stop hypothesis rejected! Proceedings of the Royal Society of London, Series B 274:735–739.

SCHMIDT, J. R., AND R. E. SHOPE. 1971. Kemerovo virus from a migrating Common Redstart of Eurasia. Acta Virologica 15:112.

SCOTT, T. W. 1988. Vertebrate host ecology. Pages 257–280 *in* The Arboviruses: Epidemiology and Ecology, vol. 1 (T. P. Monath, Ed.). CRC Press, Boca Raton, Florida.

SCOTT, T. W., G. S. BOWEN, AND T. P. MONATH. 1984. A field study of the effects of Fort Morgan virus, an arbovirus transmitted by swallow bugs, on the reproductive success of Cliff Swallows and symbiotic House Sparrows in Morgan County, Colorado, 1976. American Journal of Tropical Medicine and Hygiene 33:981–991.

SCOTT, T. W., J. D. EDMAN, L. H. LORENZ, AND J. L. HUBBARD. 1988. Effects of disease on vertebrates' ability behaviorally to repel host-seeking mosquitoes. Miscellaneous Publications of the Entomological Society of America 68:9–17.

SCOTT, T. W., AND L. H. LORENZ. 1998. Reduction of *Culiseta melanura* fitness by eastern equine encephalomyelitis virus. American Journal of Tropical Medicine and Hygiene 59:341–346.

SCOTT, T. W., L. H. LORENZ, AND J. D. EDMAN. 1990. Effects of House Sparrow age and arbovirus infection on attraction of mosquitoes. Journal of Medical Entomology 27:856–863.

SCOTT, T. W., AND S. C. WEAVER. 1989. Eastern equine encephalomyelitis virus: Epidemiology and evolution of mosquito transmission. Advances in Virus Research 37:277–328.

SCOTT, T. W., S. C. WEAVER, AND V. L. MALLAMPALLI. 1994. Evolution of mosquito-borne viruses. Pages 293–324 *in* The Evolutionary Biology of Viruses (S. S. Morse, Ed.). Raven, New York.

Sellers, R. F. 1989. Eastern equine encephalitis in Quebec and Connecticut, 1972: Introduction by infected mosquitoes on the wind? Canadian Journal of Veterinary Research 53:76–79.

Sellers, R. F., and A. R. Maarouf. 1988. Impact of climate on western equine encephalitis in Manitoba, Minnesota and North Dakota, 1980–1983. Epidemiology and Infection 101:511–535.

Sellers, R. F., and A. R. Maarouf. 1990. Trajectory analysis of winds and eastern equine encephalitis in USA, 1980–5. Epidemiology and Infection 104:329–343.

Sellers, R. F., and A. R. Maarouf. 1993. Weather factors in the prediction of western equine encephalitis epidemics in Manitoba. Epidemiology and Infection 111:373–390.

Sellers, R. F., D. E. Pedgley, and M. R. Tucker. 1982. Rift Valley fever, Egypt 1977: Disease spread by windborne insect vectors? Veterinary Record 110:73–77.

Shaman, J. 2007. Amplification due to spatial clustering in an individual-based model of mosquito-avian arbovirus transmission. Transactions of the Royal Society of Tropical Medicine and Hygiene 101:469–483.

Shirako, Y., B. Niklasson, J. M. Dalrymple, E. G. Strauss, and J. H. Strauss. 1991. Structure of the Ockelbo virus genome and its relationship to other Sindbis viruses. Virology 182:753–764.

Shope, R. 1991. Global climate change and infectious diseases. Environmental Health Perspectives 96:171–174.

Shope, R. E., A. H. Paes de Andrade, G. Bensabath, O. R. Causey, and P. S. Humphrey. 1966. The epidemiology of EEE, WEE, SLE, and Turlock viruses, with special reference to birds, in a tropical rain forest near Belem, Brazil. American Journal of Epidemiology 84:467–477.

Small, A. 1974. The Birds of California. Winchester, New York.

Smith, R. P., Jr., P. W. Rand, E. H. Lacombe, S. R. Morris, D. W. Holmes, and D. A. Caporale. 1996. Role of bird migration in the long-distance dispersal of Ixodes dammini, the vector of Lyme disease. Journal of Infectious Diseases 174:221–224.

Sooter, C. A., B. F. Howitt, and R. Gorrie. 1952. Encephalitis in the Midwest. IX. Neutralizing antibodies in wild birds of Midwestern states. Proceedings of the Society of Experimental Biology and Medicine 79:507–509.

Spence, L. P. 1980. St. Louis encephalitis in tropical America. Pages 451–471 in St. Louis Encephalitis (T. P. Monath, Ed.). American Public Health Association, Washington, D.C.

Sprance, H. E. 1981. Experimental evidence against the transovarial transmission of eastern equine encephalitis virus in Culiseta melanura. Mosquito News 41:168–173.

Stallknecht, D. E. 2007. Impediments to wildlife disease surveillance, research, and diagnostics. Current Topics in Microbiology and Immunology 315:445–461.

Stamm, D. D. 1958. Studies on the ecology of equine encephalomyelitis. American Journal of Public Health 48:328–335.

Stamm, D. D., and R. J. Newman. 1963. Evidence of southward transport of arboviruses from the U.S. by migratory birds. Annals of Microbiology 11:123–133.

Still, E., P. Monaghan, and E. Bignal. 1987. Social structuring at a communal roost of Choughs Pyrrhocorax pyrrhocorax. Ibis 129:398–403.

Stoops, A. C., K. A. Barbara, M. Indrawan, I. N. Ibrahim, W. B. Petrus, S. Wijaya, A. Farzeli, U. Antonjaya, L. W. Sin, N. Hidayatullah, and others. 2009. H5N1 surveillance in migratory birds in Java, Indonesia. Vector-Borne and Zoonotic Diseases 9:695–702.

Strauss, J. H., and E. G. Strauss. 1994. The alphaviruses: Gene expression, replication, and evolution. Microbiology Reviews 58:491–562.

Stutchbury, B. J. M., S. A. Tarof, T. Done, E. Gow, P. M. Kramer, J. Tautin, J. W. Fox, and V. Afanasyev. 2009. Tracking long-distance songbird migration by using geolocators. Science 323:896.

Sullivan, H., G. Linz, L. Clark, and M. Salman. 2006. West Nile virus antibody prevalence in Red-winged Blackbirds (Agelaius phoeniceus) from North Dakota, USA (2003–2004). Vector-Borne and Zoonotic Diseases 6:305–309.

Takahashi, M., A. Oya, M. Morita, H. Hasegawa, and S. Kudo. 1972. An ornithological study for arboviruses in migrating birds of transcontinental route to Japan. Pages 189–192 in Transcontinental Connections of Migratory Birds and Their Role in the Distribution of Arboviruses. Nauka, Academy of Sciences of the USSR, Siberian Branch, Biological Institute, Novosibirsk, Siberia (USSR).

Tatem, A. J., S. I. Hay, and D. J. Rogers. 2006. Global traffic and disease vector dispersal. Proceedings of the National Academy of Sciences USA 103:6242–6247.

Taylor, R. M., H. S. Hurlbut, T. H. Work, J. R. Kingston, and T. E. Frotingham. 1955. Sindbis virus: A newly recognized arthropod-transmitted virus. American Journal of Tropical Medicine and Hygiene 4:844–862.

Taylor, R. M., T. H. Work, H. S. Hurlbut, and F. Rizk. 1956. A study of West Nile virus in Egypt. American Journal of Tropical Medicine and Hygiene 5:579–620.

Tesh, R. B. 1984. Transovarial transmission of arboviruses in their invertebrate vectors. Pages 57–76 in Current Topics in Vector Research, vol. 2 (K. F. Harris, Ed.). Praeger, New York.

Tesh, R. B., R. Parsons, M. Siirin, Y. Randle, C. Sargent, H. Guzman, T. Wuithiranyagool, S. Higgs,

D. L. Vanlandingham, A. A. Bala, and others. 2004. Year-round West Nile virus activity, Gulf Coast region, Texas and Louisiana. Emerging Infectious Diseases 10:1649–1652.

Tesh, R. B., M. Siirin, H. Guzman, A. P. A. T. da Rosa, X. Wu, T. Duan, H. Lei, M. R. Nunes, and S.-Y. Xia. 2005. Persistent West Nile virus infection in the golden hamster: Studies on its mechanism and possible implications for other flavivirus infections. Journal of Infectious Diseases 192:287–295.

Thogmartin, W. E., F. P. Howe, F. C. James, D. H. Johnson, E. T. Reed, J. R. Sauer, and F. R. Thompson III. 2006. A review of the population estimation approach of the North American Landbird Conservation Plan. Auk 123:892–904.

Tracey, J. P., R. Woods, D. Roshier, P. West, and G. R. Saunders. 2004. The role of wild birds in the transmission of avian influenza for Australia: An ecological perspective. Emu 104:109–124.

Turell, M. J. 1988. Horizontal and vertical transmission of viruses by insect and tick vectors. Pages 127–152 *in* The Arboviruses: Epidemiology and Ecology, vol. 1 (T. P. Monath, Ed.). CRC Press, Boca Raton, Florida.

Unnasch, R. S., E. W. Cupp, and T. R. Unnasch. 2006a. Host selection and its role in transmission of arboviral encephalitides. Pages 73–89 *in* Disease Ecology: Community Structure and Pathogen Dynamics (S. K. Collinge and C. Ray, Eds.). Oxford University Press, Oxford, United Kingdom.

Unnasch, R. S., T. Sprenger, C. R. Katholi, E. W. Cupp, G. E. Hill, and T. R. Unnasch. 2006b. A dynamic transmission model of eastern equine encephalitis virus. Ecological Modeling 192:425–440.

van den Hurk, A. F., S. A. Ritchie, and J. S. Mackenzie. 2009. Ecology and geographical expansion of Japanese encephalitis virus. Annual Review of Entomology 54:17–35.

van Gils, J. A., V. J. Munster, R. Radersma, D. Liefhebber, R. A. M. Fouchier, and M. Klaassen. 2007. Hampered foraging and migratory performance in swans infected with low-pathogenic avian influenza A virus. PLoS One, e184.

Venkatesan, M., and J. L. Rasgon. 2010. Population genetic data suggest a role for mosquito-mediated dispersal of West Nile virus across the western United States. Molecular Ecology 19:1573–1584.

Ward, M. P., A. Raim, S. Yaremych-Hamer, R. Lampman, and R. J. Novak. 2006. Does the roosting behavior of birds affect transmission dynamics of West Nile virus? American Journal of Tropical Medicine and Hygiene 75:350–355.

Watson, G. E., R. E. Shope, and M. N. Kaiser. 1972. An ectoparasite and virus survey of migratory birds in the eastern Mediterranean. Pages 176–180 *in* Transcontinental Connections of Migratory Birds and Their Role in the Distribution of Arboviruses. Nauka, Academy of Sciences of the USSR, Siberian Branch, Biological Institute, Novosibirsk, Siberia (USSR).

Watts, D. M., and B. F. Eldridge. 1975. Transovarial transmission of arboviruses by mosquitoes: A review. Medicine and Biology 53:271–278.

Watts, D. M., S. Pantuwatana, G. R. DeFoliart, T. M. Yuill, and W. H. Thompson. 1973. Transovarial transmission of LaCrosse virus (California encephalitis group) in the mosquito, *Aedes triseriatus*. Science 182:1140–1141.

Watts, S. L., D. M. Fitzpatrick, and J. E. Maruniak. 2009. Blood meal identification from Florida mosquitoes (Diptera: Culicidae). Florida Entomologist 92:619–622.

Weaver, S. C. 2006. Evolutionary influences in arboviral disease. Current Topics in Microbiology and Immunology 299:285–314.

Weaver, S. C., and A. D. T. Barrett. 2004. Transmission cycles, host range, evolution and emergence of arboviral disease. Nature Reviews Microbiology 2:789–801.

Weaver, S. C., C. Ferro, R. Barrera, J. Boshell, and J.-C. Navarro. 2004. Venezuelan equine encephalitis. Annual Review of Entomology 49:141–174.

Weaver, S. C., A. Hagenbaugh, L. A. Bellew, L. Gousset, V. Mallampalli, J. J. Holland, and T. W. Scott. 1994. Evolution of alphaviruses in the eastern equine encephalomyelitis complex. Journal of Virology 68:158–169.

Weaver, S. C., W. Kang, Y. Shirako, T. Rumenapf, E. G. Strauss, and J. H. Strauss. 1997. Recombinational history and molecular evolution of western equine encephalomyelitis complex alphaviruses. Journal of Virology 71:613–623.

Weaver, S. C., and W. K. Reisen. 2010. Present and future arboviral threats. Antiviral Research 85:328–345.

Weaver, S. C., R. Rico-Hesse, and T. W. Scott. 1992. Genetic diversity and slow rates of evolution in New World alphaviruses. Current Topics in Microbiology and Immunology 176:99–117.

Weber, T. P., and N. I. Stilianakis. 2007. Ecologic immunology of avian influenza (H5N1) in migratory birds. Emerging Infectious Diseases 13:1139–1143.

Weissenböck, H., J. Kolodziejek, A. Url, H. Lussy, B. Rebel-Bauder, and N. Nowotny. 2002. Emergence of Usutu virus, an African mosquito-borne flavivirus of the Japanese encephalitis virus group, central Europe. Emerging Infectious Diseases 8:652–656.

Wellings, F. M., A. L. Lewis, and L. V. Pierce. 1972. Agents encountered during arboviral ecological studies: Tampa Bay area, Florida, 1963 to 1970. American Journal of Tropical Medicine and Hygiene 21:201–213.

Wheeler, S. S., C. M. Barker, Y. Fang, M. V. Armijos, B. D. Carroll, S. Husted, W. O. Johnson, and W. K. Reisen. 2009. Differential impact of West Nile virus on California birds. Condor 111:1–20.

WHITE, D. M., W. C. WILSON, C. D. BLAIR, AND B. J. BEATY. 2005. Studies on overwintering of bluetongue viruses in insects. Journal of General Virology 86:453–462.

WIKELSKI, M., E. M. TARLOW, A. RAIM, R. H. DIEHL, R. P. LARKIN, AND G. H. VISSER. 2003. Costs of migration in free-flying songbirds. Nature 423:704.

WILLIAMS, T. C., AND J. M. WILLIAMS. 1990. The orientation of transoceanic migrants. Pages 7–21 *in* Bird Migration: Physiology and Ecophysiology (E. Gwinner, Ed.). Springer-Verlag, Berlin.

WILLIAMS, T. C., J. M. WILLIAMS, L. C. IRELAND, AND J. M. TEAL. 1978. Estimated flight time for transatlantic autumnal migrants. American Birds 32:275–280.

WINKER, K., K. G. MCCRACKEN, D. D. GIBSON, C. L. PRUETT, R. MEIER, F. HUETTMANN, M. WEGE, I. V. KULIKOVA, Y. N. ZHURAVLEV, M. L. PERDUE, AND OTHERS. 2007. Movements of birds and avian influenza from Asia into Alaska. Emerging Infectious Diseases 13:547–552.

WOODALL, J. P., T. E. LOVEJOY, F. C. NOVAES, A. H. ANDRADE, AND G. BENSABATH. 1972. Some data on the role of birds in arbovirus ecology in the Amazon basin. Pages 214–218 *in* Transcontinental Connections of Migratory Birds and Their Role in the Distribution of Arboviruses. Nauka, Academy of Sciences of the USSR, Siberian Branch, Biological Institute, Novosibirsk, Siberia (USSR).

WORK, T. H., H. S. HURLBUT, AND R. M. TAYLOR. 1955. Indigenous wild birds of the Nile delta as potential West Nile virus circulating reservoirs. American Journal of Tropical Medicine and Hygiene 4:872–888.

WORK, T. H., AND R. D. LORD. 1972. Trans-Gulf migrants and the epizootology of arboviruses in North America. Pages 207–213 *in* Transcontinental Connections of Migratory Birds and Their Role in the Distribution of Arboviruses. Nauka, Academy of Sciences of the USSR, Siberian Branch, Biological Institute, Novosibirsk, Siberia (USSR).

WRIGHT, S. A., D. A. LEMENAGER, J. R. TUCKER, M. V. ARMIJOS, AND S. A. YAMAMOTO. 2006. An avian contribution to the presence of *Ixodes pacificus* (Acari: Ixodidae) and *Borrelia burgdorferi* on the Sutter Buttes of California. Journal of Medical Entomology 43:368–374.

YAREMYCH, S. A., R. J. NOVAK, A. J. RAIM, P. C. MANKIN, AND R. E. WARNER. 2004. Home range and habitat use by American Crows in relation to transmission of West Nile virus. Wilson Bulletin 116:232–239.

YASUE, M., C. J. FEARE, L. BENNUN, AND W. FIEDLER. 2006. The epidemiology of H5N1 avian influenza in wild birds: Why we need better ecological data. BioScience 56:923–929.

YOM-TOV, Y. 1979. The disadvantage of low positions in colonial roosts: An experiment to test the effect of droppings on plumage quality. Ibis 121:331–333.

YOUNG, D. S., L. D. KRAMER, J. G. MAFFEI, R. J. DUSEK, P. B. BACKENSON, C. N. MORES, K. A. BERNARD, AND G. D. EBEL. 2008. Molecular epidemiology of eastern equine encephalitis virus, New York. Emerging Infectious Diseases 14:454–460.

YUILL, T. M. 1986. The ecology of tropical arthropod-borne viruses. Annual Review of Ecology and Systematics 17:189–219.

YUNKER, C. E. 1975. Tick-borne viruses associated with seabirds in North America and related islands. Medical Biology 53:302–311.

YUNKER, C. E., C. M. CLIFFORD, L. A. THOMAS, J. CORY, AND J. E. GEORGE. 1972. Isolation of viruses from swallow ticks, *Argas cooleyi*, in the southwestern United States. Acta Virologica 16:415–421.

ZELL, R. 2004. Global climate change and the emergence/re-emergence of infectious diseases. International Journal of Medical Microbiology 293 (Supplement 37):16–26.

ZELLER, H. G., AND I. SCHUFFENECKER. 2004. West Nile virus: An overview of its spread in Europe and the Mediterranean Basin in contrast to its spread in the Americas. European Journal of Clinical Microbiology and Infectious Diseases 23:147–156.

ZHONG, H., Z. YAN, F. JONES, AND C. BROCK. 2003. Ecological analysis of mosquito light trap collections from west central Florida. Environmental Entomology 32:807–815.

Shorebird Ecology, Conservation, and Management

Mark A. Colwell

"Anyone seeking a current and comprehensive reference on the ecology and conservation of shorebirds will find Colwell's effort outstanding. It provides a sorely needed companion to the numerous identification guides that have been published recently for this diverse and fascinating group of birds."

—**Robert Gill, Project Leader, Shorebird Research, USGS**

$60.00 cloth

Encyclopedia of Biological Invasions

Daniel Simberloff and Marcel Rejmánek, Editors

"An extraordinarily useful and authoritative compilation of our current knowledge on biological invasions…. Well-organized, beautifully illustrated, and thoughtfully produced."

—**Harold Mooney, Stanford University**

Encyclopedias of the Natural World
$95.00 cloth

Top 100 Birding Sites of the World

Dominic Couzens

"For pre-trip planning (and excitement building) and post-trip remembrance…. Gorgeous color photographs."—***Booklist***

$45.00 cloth

Ornithological Monographs also available

On the Origin of Species Through Heteropatric Differentiation

A Review and a Model of Speciation in Migratory Animals

Kevin Winker

Ornithological Monographs, 69
$19.95 paper

Thomas R. Howell's Checklist of the Birds of Nicaragua as of 1993

Juan C. Martinez-Sanchez and Tom Will, Editors

Ornithological Monographs, 68
$24.95 paper

Avian Subspecies

Kevin Winker and Susan M. Haig, Editors

Ornithological Monographs, 67
$40.00 paper

Cladistics and the Origin of Birds

A Review and Two New Analyses

Frances C. James and John A. Pourtless IV

Ornithological Monographs, 66
$19.95 paper

Reproduction and Immune Homeostastis in a Long-lived Seabird, the Nazca Booby (*Sula granti*)

Victor Apanius, Mark A. Westbrock, and David J. Anderson

Ornithological Monographs, 65
$19.95 paper

Forthcoming In the *Studies for Avian Biology* series

Greater Sage-Grouse

Ecology and Conservation of a Landscape Species and Its Habitats

Steven T. Knick and John W. Connelly, Editors

$95.00 cloth

Ecology, Conservation, and Management of Grouse

Brett K. Sandercock, Kathy Martin, and Gernot Segelbacher, Editors

$70.00 cloth

Boreal Birds of North America

A Hemispheric View of Their Conservation Links and Significance

Jeffrey V. Wells, Editor

$39.95 cloth

Population Demography of Northern Spotted Owls

Eric D. Forsman, et al.

$39.95 cloth

At bookstores or order www.ucpress.edu

UNIVERSITY OF CALIFORNIA PRESS

UNIVERSITY OF CALIFORNIA PRESS

JOURNALS + DIGITAL PUBLISHING

Now it's easier than ever to

Get Reprints & Copyright Permission

Go to **ucpressjournals.com**, locate the content on Caliber that interests you, and click on "reprints and permissions" to:

◆〉 Order Reprints

– Standard or Custom,
in black and white or color

◆〉 Use Information

– Instant Permission to Reuse Articles,
Tables and Graphs

◆〉 Republish Content

– In Journals, Newsletters,
Anthologies, etc.

rightslink.copyright.com

Help Feather Our Nest

Keep AOU in the forefront of professional ornithology worldwide: Include our society in your will

A commitment to AOU through your estate can support the leadership, ideals, and programs we all share as AOU members:

- A professional journal with an outstanding citation index
- Exceptional annual meetings and the opportunity to gather with colleagues
- Travel and research funds for students

Modify your will to benefit AOU. It's simple and the benefits are enormous:

- You can make a gift to AOU without affecting your financial security.
- Tax deductions may help avoid estate capital gains taxes.
- A bequest can be designated to benefit an AOU program of your choice.
- A gift to the Endowment Fund will benefit the AOU in perpetuity.

To include AOU in your estate plans, simply contact your attorney and ask him or her to add the following language to your will:

I hereby give, devise, and bequeath to the American Ornithologists' Union, a scientific society based at 1313 Dolley Madison Blvd., McLean, VA 22101 {_____ dollars} or {_____ percent or all of the rest, residue, and remainder of my own estate} for its {general purposes, AOU's Endowment Fund, or_______ program}.

For more information, please contact Betty Anne Schreiber, AOU Development Committee, at 828-645-9704 or SchreiberE@aol.com

Order from: University of California Press, Journals and Digital Publishing, www.ucpressjournals.com, or e-mail customerservice@ucpressjournals.com.

No. 65. *Reproduction and Immune Homeostasis in a Long-lived Seabird, the Nazca Booby (Sula granti).* V. Apanius, M. A. Westbrock, and D. J. Anderson. viii + 46 pp. 2008. $19.95.

No. 66. *Cladistics and the Origin of Birds: A Review and Two New Analyses.* Frances C. James and John A. Pourtless IV. viii + 78 pp. 2009. $19.95.

No. 67. *Avian Subspecies.* Kevin Winker and Susan M. Haig, eds. viii + 200 pp. 2010. $24.95.

No. 68. *Thomas R. Howell's Check-list of the Birds of Nicaragua as of 1993.* Juan C. Martínez-Sánchez and Tom Will, Eds. xvi + 108 pp. 2010. $24.95.

No. 69. *On the Origin of Species Through Heteropatric Differentiation: A Review and a Model of Speciation in Migratory Animals.* Kevin Winker. viii + 30 pp. 2010. $19.95.

No. 70. *Avifauna of the Eastern Himalayas and Southeastern sub-Himalayan Mountains: Center of Endemism or Many Species in Marginal Habitats?* Swen C. Renner and John H. Rappole, eds. viii + 166 pp. 2011. $24.95.